The Neuroscience of Business

Series Editors
Peter Chadwick
Ideas For Leaders
London, UK

Roderick Millar
Ideas For Leaders
Edinburgh, UK

Neuroscience is changing our understanding of how the human brain works and how and why people behave the way they do. Properly understood, many of these insights could lead to profound changes in the way businesses interact with their employees and customers. The problem is that, until now, most of this research has been published in specialist journals and has not made its way to managers' desks. At the same time, however, business leaders and managers are faced with a plethora of extravagant claims based on misunderstood, or exaggerated, neuroscientific research.

Palgrave's The Neuroscience of Business series seeks to bridge the gap between rigorous science and the practical needs of business. For the first time this series will describe the practical managerial applications of this science in an accessible, but in-depth, way that is firmly underpinned by a clear explanation of the science behind the management actions proposed.

More information about this series at
http://www.palgrave.com/gp/series/14428

Kate Lanz · Paul Brown

All the Brains in the Business

The Engendered Brain
in the 21st Century Organisation

palgrave
macmillan

Kate Lanz
Bedford, UK

Paul Brown
Vientiane, Laos

The Neuroscience of Business
ISBN 978-3-030-22152-2   ISBN 978-3-030-22153-9   (eBook)
https://doi.org/10.1007/978-3-030-22153-9

This Palgrave Macmillan imprint is published by the registered company Springer Nature Switzerland AG
The registered company address is: Gewerbestrasse 11, 6330 Cham, Switzerland

Preface

It promised to be neither the best of days nor the worst of days: just another early morning in the plastic oasis of an airport transit lounge. Newspapers were ready and waiting. There was an English breakfast tea too. That seemed a promise of better things to come after a long-haul flight and a destination needing only one last flight.

Then slowly, as the sky lightened, it dawned also that the day was 9/11 and, it so happened, a Tuesday, just as had been the case seventeen years before. Memories of watching the horror of those events live on television in the curious tranquility of an early afternoon, sitting in a shaded drawing room overlooking a garden in the peace of the English west countryside, flooded in; remembering too that a trainee journalist son had planned to be somewhere downtown in New York that morning. For thirty-six hours no certainty of his fate was known.

Surprisingly there was no mention of the twin towers anniversary in the first paper that had come to hand, the *Daily Telegraph*. What was consuming editors and their journalists then? Slowly, reflectively, it became apparent that another kind of American tragedy was being acted out, this time a social one, and spreading virally across the western world.

* * *

The previous Saturday in the U.S. Open Final, Serena Williams had raged on court at what she considered to be umpiring decisions prejudiced against her. Whatever the merits of her claim and actions, or her self-justification in saying: 'I have a daughter and I stand for what's right', they generated many columns of reportage. And much spin-off comment too.

'Serena's rage was magnificent and a lesson for all', wrote Celia Walden on page 20. 'I say it was both a triumph and a reminder of why women need to get better at rage'.

Really?, the early-morning brain wondered. Has any great society ever made a virtue of feminine rage? There is righteous indignation; and then there is rage. Getting better at it is not a cultural proposition widely held. Not yet, anyway. Should it be?

The balance of power between men and women is not a new topic. Nearly two-thousand-five-hundred years ago Aristophanes was exploring it in *Lysistrata*. Wikipedia[1] has an elegant entry on the play. It says, in part:

Lysistrata … is an extraordinary woman with a large sense of individual and social responsibility. She has convened a meeting of women from various Greek city-states that are at war with each other (there is no mention of how she managed this feat) and, very soon after confiding in her friend about her concerns for the female sex, the women begin arriving.

With support from the Spartan, Lampito, Lysistrata persuades the other women to withhold sexual privileges from their menfolk as a means of forcing them to end the interminable Peloponnesian War. The women are very reluctant, but the deal is sealed with a solemn oath around a wine bowl, Lysistrata choosing the words and Calonice repeating them on behalf of the other women. It is a long and detailed oath, in which the women abjure all their sexual pleasures, including the Lioness on the Cheese Grater (a sexual position).

Soon after the oath is finished, a cry of triumph is heard from the nearby Acropolis—the old women of Athens have seized control of it at Lysistrata's instigation, since it holds the state treasury, without which the men cannot long continue to fund their war. Lampito goes off to spread the word of revolt, and the other women retreat behind the barred gates of the Acropolis to await the men's response.

So there are the women of classical times using all the resources available to them. Maybe Thurber had the play in mind when he observed that 'Nobody will ever win the battle of the sexes. There's just too much fraternising with the enemy'.[2]

---

[1]https://en.wikipedia.org/wiki/Lysistrata (accessed 17 October 2018).

[2]There is some doubt as to the original author of this fine thought, but Thurber is a strong candidate and it sounds very Thurber-esque.

Celia Walden's cry for raging women was preceded in the *Telegraph* by the banner that a sub-editor had raised for Julia Hartley-Brewer's column four pages earlier. 'Let us be rude, or we'll all go to hell in a handbasket' said the headline. She was in fact making a reasoned case for the full use of expressive language, including metaphor, because a policeman had been disciplined for using the phrase 'Whiter than white'. And she quoted with approval Rowan Atkinson's question about what happened to our right to offend? But the sub-editing had slipped into the convenient modern canard that stridency is a virtue. How many women in organisations have attended training courses to help them become more assertive in Board meetings where the very clear implied message is: *Be more like men.*

Is that really the future?

Nearly six decades of active feminism seem, at this point of social history, to have got dead-ended in the argument for equality—a pursuit and demand that in a dynamic society can have no ending: ever. It's a losers' catch-up game. Perhaps the origins of the demand come from the collective American non-consciousness created by the Declaration of Independence when, poorly worked out in practice though it was, equality was enshrined as being among the unalienable rights of existence: of 'all men', at least.

This book has been written with the underlying intention to help avert the social tragedy that, put simply, is about the widespread loss in western societies of trust between women and men and the consequent breakdown of mannerly relationships between them. Put more specifically, it is about the loss of cultural delight in the differences between women and men. We mourn that loss. More purposefully, we want women to claim what is theirs *in their own right*, not by using men as the standard of reference for some ungraspable equality. Being *co-equal* is a different proposition from equality as it implies a coming together, not a catching up.

So we are starting the argument for maximising women's claim to *the value of the difference*. If women are to be valued for who they are, not whether they have attained some state of comparative manhood, then new values will have to come into play too. We want to show that modern scholarship and applied neuroscience are beginning to understand the differences. And they have, we believe, huge ramifications for the future of sustainable organisations.

* * *

Our essential aim is to help shift the debate about women in the workplace into something seriously purposeful and achievable; and in consequence to get out of the cul-de-sac in which modern feminism finds itself and, frustratingly for both sides, blames men for the way things are. For the demand for equality has led to a perverse and unsustainable illogical conclusion. From the assertion that if women can do anything men can do (quite right) then men and women must be the same (completely wrong).

Women can arrive at the same outcomes that men can create, or they might arrive somewhere else, but they would do it differently. That is our essential proposition and continuous frame of reference.

The achievement of sixty years of feminism in creating the huge cultural shift that it has, laying claim to women being in the world—whatever world they choose to be in—as individuals in their own right, is extraordinary. Although there are lots of battles still to be fought, the war has been won—at least in the West. In the world of organisations, however, part of the price paid for that achievement has been that in so many organisations women not only find themselves having to be the best men they can be but, when it all gets too inconvenient for process-driven systems, then they can be shunted off into being part of the organisation's diversity agenda.

That really won't do either, will it?

And this book is not just about men and women. It is very much about the fact that brains themselves can be sexually typed. Just because one person is observably female and another male, and variably placed on a feminine/masculine dimension, we cannot assume that the brain either possesses is of a similar sex. The brain itself within the person may be more feminine or masculine. When this is understood, then the possibilities for a deeper understanding of what the organisational implications of the differences are becomes even more exciting, we suggest.

Neuroscience is making us re-think the fundamentals of human behaviour. Twentieth-century psychology had lots of theories about human behaviour but no certainties. Psychologists could not agree among themselves, and failed to provide a firm scientific basis for either psychiatry (disturbances of the human condition) or HR (ways of disturbing the human condition).[3] The neurosciences of the twenty-first century are telling us that individuals are not, primarily, psychological systems but physical systems: and that individuality is created by the complexity of how experience attached to genetics creates a different brain for everyone even if the unit parts are essentially the same.

There are only twenty-six letters in the English alphabet, yet every book that is written re-arranges (within an agreed rule base) the letters that form

---

[3]A joke. Ed.

the words that create the plethora of unique books. So it is with humans. The component parts of each individual's brain—an extraordinary neuro-chemical factory being got ready for its adult life purposes through twenty-four years of development—are systematically organised by the way emotions (e-motions) create pathways in the brain—a process that starts from month four of uterine life.

So, like the rest of the physical world, we human beings are essentially *energy* systems. What we call our psychology is in fact a very complex set of unique algorithms, created by the emotional system that drives each of us into being our—Self and trying to maintain that Self through all the vicissitudes of living a life.

But it seems most likely that feminine and masculine energy are different; are organised biologically for different purposes; and manifest themselves in adult life as forces that, in context, may have very different and complementary value even if both male and female share (social) values that are identical.

What we have set ourselves the task of getting into print is to answer the question: *What happens when feminine energy begins to be understood and properly valued in the organisational context?* That's the organisational conundrum we want to solve, asking also on the way: *And why does it have to be a masculine context?*

If we were to sum it up in one deceptively simple statement, what we are re-formulating is not an argument for equality but, for organisational advantage, the emergent quality of women: *e-quality*, we have started to call it. Even, sometimes, s/he-quality: an emergence of a shared understanding, as between men and women, of the quality of the energy that each bring to corporate endeavour; and of the delight in value that, together, men and women create in a way that neither can create alone.

Kate Lanz, Bedford, England + Paul Brown, Vientiane, Laos PDR. April 2019.

PS: While in preparing this book we have each commented extensively on what the other wrote, we thought it would be in the spirit of female and male to keep them separate as chapters as well as combined as a book.

A generous quirk of the publisher has allowed that to happen.

Bedford, UK   Kate Lanz
Vientiane, Laos   Paul Brown

Acknowledgements

Thanks to my loving family for all their support
especially to Harry and George
Nen and Nic
and my dear friend Paul

To Ketta,
for all her quiet support
and with love.

About This Book

What You Need to Know About the Brain at Work

Before we start on our journey of understanding how to get the best of all the brains in the business, it is important to share some basic facts about the brain. It is also important to note that neuroscience is still in its infancy. It is far more likely that there is far more about the brain that is not known than is known. The explanation that follows is a simplification of the complex neurobiological processes that happen in the brain. That said, there is now a sufficient body of scientific evidence to allow our understanding of human behaviour to begin shift from a descriptive model to an explanatory one. The science is there and we want to share an understanding of its application in business with you in this book. To begin the journey, here is a simple overview of how our brains work.

The Bottom-Up Triune Brain

The Reptilian Brain

The brain functions in a bottom-up fashion. This is in line with millions of years of human evolutionary history. The oldest part of the brain, inherited from our reptilian ancestors is the brain stem, sitting at the base of the brain atop the spinal column. Its colloquial name is the 'reptilian brain'. It is some 360 million years old and manages many of the autonomic functions of the body—breathing, heart rate, digestive system…and so on… When someone is declared 'brain dead' this is often the only part of the brain still functioning. The person is technically alive but has no consciousness. This early part of the brain served our ancient reptilian ancestors who reproduced by laying eggs. There was no relationship needed between the parent and offspring that would impact the survival of the babies.

The Emotional Limbic Brain

Some 65 or so million years ago the first mammals began to appear, giving birth to live young. For the offspring of mammals to survive there had to be a relationship with the parent. Neuroscience suggests[4] that the limbic brain evolved at this time with one of its primary functions being to enable the mammals to connect emotionally with each other in the pursuit of survival. In Homo sapiens the limbic system is still the brain's seat of emotions, primal survival emotions that would have helped to ensure survival for our ancient ancestors on the savannah. All input from the five senses is filtered first and foremost by the limbic system. The limbic system's read on our environment is emotional—not cognitive. Its main concern is for our survival and it responds with primal emotion within 85 milliseconds to a change in our environment with a view to keeping us safe. Within the limbic system the amygdala is the guardhouse of the brain and is scanning for threat and danger all the time. Perhaps we all recognise that before we have had time to cognitively make sense of a situation the amygdala has already generated a neurochemical response designed to protect us, our mother's advice to count to ten before reacting is based soundly in modern neuroscience.

The Brain's CEO—The Prefrontal Cortex

The third part of the brain to evolve is the cortex. The prefrontal cortex (PFC) is the CEO of our brains. It is the part of the brain that makes rational sense of the world, adds meaning to our experience, enables complex decision-making. It is the part of our brain that makes us smart sophisticated beings capable of creating the complex global societies we live in. The PFC in evolutionary terms is a toddler. It is some 220,000 years old.

---

[4]MacLean, P. D. (1973). *A Triune Concept of Brain and Behavior.* University of Toronto.

The PFC comes on stream about 250 milliseconds after a stimulus—in time to try to make sense of sensory information our limbic system has already responded to.

Emotional Before (Cognitively) Rational

This is a really important neurobiological fact for leaders in business to understand. Our emotional response is much faster than our rational thought—three times as fast. A friend of Kate's working on the human genome project once exclaimed to her at a Christmas party 'the thing is Kate, our brains have not evolved since we came out of the caves. We have a stone age brain in the modern world we have created'. There is some truth in this notion. Our immediate response to any situation is inevitable governed by our limbic system which is the seat of primal survival emotions, designed above all else to keep us alive. Any subsequent response is very likely to be influenced by this initial stance.

Primary Colours of Emotion

An artist creates an array of different nuanced colours from a palette of three basic primary colours—so it is with the emotional system. Neuroscience generally agrees that there are eight primal emotions that show up across cultures. Five of these emotions exist to ensure our survival. These are **Fear**, **Anger**, **Disgust Sadness and Shame**. These five emotions would have kept us alive on the plains inhabited by our ancestors. Fear tells us to run, fight or hide; Anger to tell us to defend and kill; Disgust to indicate not to eat something that might be toxic to us; Sadness to tell us to whom we should attach ourselves, our primary caregivers ensuring our survival, and finally shame to keep us behaving well enough for our continued acceptance as part of the tribe. All of these emotions will trigger the brain to generate (among others) the neurotransmitters adrenaline and cortisol to prime us for the appropriate action.

Surprise!

The sixth emotion is surprise—a potentiator to tip us into a survive response if needed. Or alternatively if the surprise is a positive one and we are faced with friend not foe, then to tip us into thrive.

Thrive

There are two primary thrive emotions. These emotions do not ensure our survival in the first instance. They will not keep us alive, but they will help us do well and help our brains and emotional systems to develop in a healthy way. The first primal thrive emotion is love—the same basic emotion as trust. The second is joy. The neurochemical hallmarks of thrive are oxytocin the bonding hormone, serotonin the happiness neurotransmitter and dopamine the reward neurotransmitter. In order for learning to occur, the brain needs the right amount of dopamine in the right place in the brain at the right time.

Survive Versus Thrive and the Relevance for Business

Brains that are in a thrive state are more creative, learn better and have more energy available to the cortex for complex problem-solving and are more agile in taking action. The brain is a highly complex connected network. Our brains move through emotional states very very fast. And it is true to say and we will evidence the business impact later in this book that brains that are enabled to be in a thrive state (emotionally and neurochemically) are more effective brains in the workplace.

Read on.

Contents

List of Figures

1

Brain Sex and Biological Sex

Kate says:

The Case for Creating Optimal Brain Conditions

During the financial crisis, I was coaching an extremely experienced, bright, delightful gentleman who had been hired to help the company in question to get themselves out of the trouble they were in. In our second session, he looked at me and said, 'Do you know what? I have to switch myself *off* in the morning just to survive the day here and I switch myself back on when I get home and see my kids' faces as I open the door'. I was horrified—what an awful way to spend one's life. The company in question were paying him a substantial sum to help them get back on track. I found myself asking, 'Why would you pay that brain that much money and then create the conditions that switch it off like that?' The return on investment on his brain power was appalling.

So began my research: what does it take to create the conditions for optimal brain performance in the workplace?

Optimal brain energy flow, such that the powerful, decision-making part of the brain, the prefrontal cortex, is facilitated to do its best work, is good for business. Creating innovative solutions and having the agility to move from strategy to execution fast are becoming increasingly important in the modern, global economic environment. It's the people in the business— their brains—that either enhance innovation and agile execution or slow it

© The Author(s) 2020
K. Lanz and P. Brown, *All the Brains in the Business*, The Neuroscience of Business,
https://doi.org/10.1007/978-3-030-22153-9_1

down. Brains that are enabled to thrive to work optimally are brains that like to come to work and they produce the neurochemistry for performance.

The Sex of Your Brain

It became very clear, very quickly that in order to answer the question, 'what does it take to create the conditions for optimal brain performance in the workplace?', biological sex differences simply could not be ignored. There *are* differences between a male and female brain (Brizendine 2007a)—and knowing how to access brain gender difference is a source of competitive advantage. However, it is not as simple as binary male–female brain differences. The apparent sex of the person does not define the functioning of their brain.

Each of our brains is different. Modern neuroscience suggests that we can characterise an individual brain within a three-dimensional space consisting of brain structure, neural connectivity and hormone levels. A combination of our unique nature and nurture determines where our brain fits within this space. On average, male and female brains exist in different regions of this space—there are measurable differences in structure, connectivity and hormone levels. Understanding where an individual sits within this space, which mixture of female–male characteristics they possess, is the basis for helping them to thrive. In our work, we are trying to help leaders understand the diversity of brains that they have within their businesses in order to create the conditions to get the best from them.

Added to the important acknowledgement of differences between the male and female brain, it is a truly fascinating and hardly recognised fact that the sex of your brain may not be the same as the sex of your body (Moir and Jessel 1989, p. 50).

How is that? In short, it is because the way your brain develops and the trillions of synaptic connections that make us who we are determined by the unique blend of nature and nurture. There are some fundamental biological facts which determine our biological sex. These combine with the influences of our environment throughout our lives, but notably in the womb and the first two years (Schore 2001) determine how our genetic blueprint gets expressed and shapes the neural pathways that make us—us. Our unique combination of nature and nurture determines the sex of our brain. Modern neuroscience is demonstrating that the majority of us have a 'mosaic' brain (Joel et al. 2015) that is a mix of both male and female characteristics. Our brain patterning within this space is as unique to each of us as is our fingerprint.

Biological sex drives different behaviours, yet within these differences lies a far greater subtlety and source of significant performance potential for business. Knowing how to access the total brain sex diversity in a business is a new source of serious competitive advantage. By enabling different sexed brains to be allowed to function in a **thrive** rather than a **survive** mode, businesses can truly tap into all the brains in the business. Company-specific research is showing an increased latent productivity of between 30 and 50% demonstrating that often in modern work cultures there is significant brain power wastage occurring. This book aims to help fix that.

Before we take a look at some of the subtle differences that make everyone's brain unique it is necessary to take a look at the overall big picture; what, on average, are some of the differences between the male and female brain?

Brain and Biological Sex—Key Differences Between the Male and Female Brain

There is a currently a naïve interpretation of feminism which seeks to assert the exact equivalence of men of women. It states that there are no significant differences between male and female brains and that social sexual stereotypes bias the way brains develop. While our social environment has some impact on our sense of self this claim is simply fake news; the science is unambiguous and it is important to understand a little about how these measurements have been made if we are to benefit from the latest discoveries.

Sex differences between male and female brains show up in the three main biological parameters that define the space within which all brain types exist. These are: particular structures within the brain characterised by volume and density differences (Cohen 2014); blood flow and thus subsequent connectivity between different brain regions (Ingalhalikar et al. 2014) and quantities and potency of certain hormones (Brizendene 2007b). Let's have a brief look at each of these to get us started on our journey of understanding brain sex difference.

Structural Differences Between the Male and Female Brain

In 2014, in the first meta-analysis of its kind into sex differences in the brain, a team of international neuroscientists (Joel et al. 2015; Cohen 2014) established that there are indeed specific structural differences between the male and female brain. On average, men's brains have larger total volume than females, though both contain the same number of neurons. Research also shows no difference in IQ between men and women (Gilmore et al. 2007; Willerman et al. 1991; Witelson et al. 2006).

The 2014 meta-analysis showed that in terms of volume and density in particular brain structures, there are differences in the limbic and language systems between men and women—the limbic system is the area of the brain responsible, in general, for emotion and memory. An interesting asymmetry was found in sex differences between the structures in question. On average men showed higher densities, mostly limited to the left side of the limbic system, whereas women showed larger limbic volumes in the right hemisphere which is more related to language, discovery and emotion. Women also showed larger specific limbic structures such as the right insular cortex and anterior cingulate gyrus—the parts of the brain that are responsible for emotion and worrying about what is going on for people.

So—what do these differences mean? In isolation, not much. However, within the context of the other major differences between the male and female brain and an understanding of brain sex as opposed to biological sex, important nuances relevant to the workplace emerge.

Blood Flow and Connectivity Differences Between the Male and Female Brain

The second major area of difference between the male and female brain is in the blood flow and neural connectivity across the network of neurons that make up the brain. In a groundbreaking study from 2013 at the University of Pennsylvania, Associate Professor Regini Verma (Ingalhalikar et al. 2014) found greater neural connectivity from front to back and *within* each hemisphere in males and greater connectivity *between* hemispheres in females. For men, this suggests that their brains are structured to enable greater connectivity between perception and coordinated action. In female brains, the greater left–right hemisphere connectivity suggests greater communication between the analytical and intuition (Fig. 1.1).

'These maps show us a stark difference–and complementarity–in the architecture of the human brain that helps provide a potential neural basis as to why men excel at certain tasks, and women at others', said Verma.

For instance, on average men are likely to be better at a single task (cycling, navigating) where women's brains seem to be more suited to superior memory and social cognition skills, making them more equipped for multi-perspective taking and creating solutions that work for a group. We investigate how these differences can be harnessed as a source of competitive advantage in the work-place later.

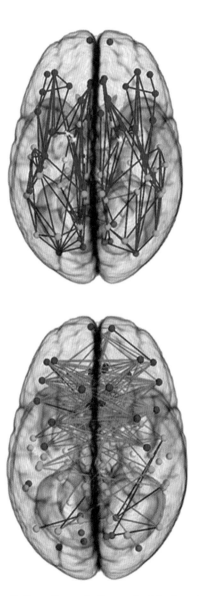

Fig. 1.1 The intra-connectivity patterns in the male (blue) and inter-connectivity patterns in the female (orange) brains (*Source* Ingalhalikar et al., *PNAS* 2014 January, 111 (2), 823–828. https://doi.org/10.1073/pnas.1316909110)

Hormones

The third major difference between the male and female brain is the impact of hormones on the brain and behaviour. There are some key sex hormones which influence how we feel and what we do. Men's bodies produce higher levels of testosterone than women's bodies. A man produces between 5 and 20 times more testosterone than a woman (Torjesen and Sandnes 2004; Southern et al. 1965, 1967; Dabbs and Dabbs 2000). Men produce testosterone from the testes—in the female body testosterone is produced by the adrenal gland and is less potent than that in men. The impact of the testosterone drives more male-type behaviours, such as competition, focusing on the importance of hierarchy and defending one's turf. Generally, women focus much less on these things and the difference in testosterone levels is one of the key reasons why.

The female body produces more oxytocin than the male body by a factor of between 5 and 20 times. Oxytocin is the bonding or trust hormone. These higher levels of oxytocin make women more likely to collaborate and connect rather than to compete. It is not just as simple as that, of course, and there are highly competitive women who care greatly about their place in the order of things. However, it is well evidenced that females tend more to nurturing, collaborative behaviours than men and, in general, care less about their place in the pecking order. Testosterone and oxytocin level differences play a large part in explaining this fact.

This key difference between testosterone and oxytocin levels in men and women, along with potency levels, accounts for a great deal of the difference in male and female behaviour in the workplace. In short and in general, women are more likely to **collaborate** *and not compete interpersonally* as their default position and men more likely to **compete** rather than collaborate. The recent research into the patriarchy paradox (Mac Giolla and Kajonius 2018) supports this distinction. This paradox refers to the fact that in a country where there is more gender equality (by law and in the culture) there is a greater difference that shows up in the way men and women think. Both behaviours are necessary to win in business.

In my experience as a leadership coach of over 20 years and a female executive in large corporates before that, it is the male tendency to compete at work which predominates in shaping modern work cultures and practices. Countless times I have come across development programmes for women which are in essence trying to get them to be the best men they can be. This is entirely missing the point, losing access to a significant source of competitive advantage. Understanding and accessing the deep essence of the female brain is available to smart businesses that create the work practices and cultures that support the best of all the brains in the business. It is becoming increasingly apparent in the modern business world that collaboration is key. Organisations need to become more effective at recognising, enabling and harnessing the collaborative power of the female brain if they want to maximise their agility.

A Look at the Impact of Nature on the Developing Brain

There are two key phases in our physical brain development that are important to understand as we move on to consider the impact of social conditioning on the observable physical differences in the brain introduced above. These differences occur in the brain in utero and also during puberty. Then there is an important third and on-going influence on the way our brain develops—the influence that key relationships and experiences shape our brains in becoming 'us' as we grow up.

The Brain *in Utero*

We will start with a brief but important look at what happens at the beginning, with the big part that Mother Nature plays. A fertilised egg in the womb contains (generally speaking) either XX chromosomes (female) or XY chromosomes (male) determining the biological sex of the developing foetus. The placenta, which acts a giant gland between the baby and its mother, also has a sex because it comes from the same cells as the baby-to-be (Mckay 2018, p. 591). It is central to sex differences in the development of the foetus.

For the first seven weeks in utero, the sex of the embryo cannot be distinguished. The basic blueprint for the brain, however, is female (Brizendene 2007a, p. 13). Professor Margaret McCarthy at the University of Maryland school of medicine summarises this by saying that 'the developing mammalian brain is destined for a female phenotype'. In the absence of a Y chromosome, the brain will, therefore, continue to develop into a female brain. When the embryo is between 6 and 8 weeks old and in the presence of the Y chromosome, a gene that is responsible for switching on the production of the testes is triggered, switching on a series of genes to define the male foetus and at the same time switching off a series of genes that prevent the further development of the female foetus. Jenny Graves from La Trobe University explains, 'Male hormones such as testosterone are synthesized by the embryonic testis and have far flung effects all over the developing body. Androgens (such as testosterone) turn on hundreds maybe thousands of genes that determine male genitalia, male growth, hair, voice and elements of behaviour' (McKay 2018, p. 620).

We have evolved to reproduce. Mother Nature is selfish and successful reproduction is her entire ambition for us. The sex hormones in the brain feminise or masculinise the parts of the brain that control reproduction depending on whether the foetus is destined to be female with XX or male with XY—chromosomes. There is a period in utero when the developing brain is extremely sensitive to the sex hormones. The dominant influence at this time is from testosterone. Under its influence, the reproductive regions of the brain become more masculinised. In the absence of testosterone these same brain areas go on developing in being feminised, the default position of the embryonic brain. The brain's reproductive areas have their male or female blue print laid down at this sensitive time. There's another surge of hormones that impacts brain development at puberty.

The First Two Years of the Developing Brain—The Beginnings of the Impact of Nurture on Nature

The last thirty years of developmental neuroscientific research has demonstrated a dynamic relationship between the switching on and off of the genes in the developing brain, as just described, and the social experience of the infant. The brain doubles in size during the individual's first year. This period is critical for healthy brain development as there is significant brain plasticity, whereby the brain changes and moulds itself according to its experience, during this period. Young children's brains do not learn or develop fully without continuous relationship connection with others. Healthy attachment relationships with primary caregivers provide the nurture input that an infant brain needs to develop in a positive way. Interacting with interested, supportive parents and being stimulated by interesting things and seeing new places is vital for healthy brain development in the infant and toddler and enables a young brain to thrive (Schore 2001).

The right hemisphere of the brain develops before the left hemisphere. The right brain is dominant in regulating emotion (US National Library of Medicine 1997; Schore 2000), managing internal states and handling our social interactions. In infancy, it is the to-and-fro between mother and baby that supports the baby to regulate its emotions and learn gradually how to do this for itself—self-soothing. These early interactions lay the foundation for self-regulation and management of emotion during later life. This is known as attachment patterning (Schore 2000).

The very early interaction with parents and caregivers shapes the child's emotional and relation attachment. This profoundly influences how the child will form relationships and interact with others in the future. These early attachment patterns, during childhood, adolescence and young adulthood shape who we are and how we show up as leaders many years later. It is vital to understand these patterns in ourselves and others in order to create the conditions for optimal brain performance at work. In the same way that a mother regulates the emotional state of the child, the organisational culture and work practices regulate, in a positive or negative way, the brains of all the people in the business. A dilemma is that it is much easier to create processes and cultures that limit the capacity of the expensive brains that come to work every day than to create the conditions in which they are enabled to give of their best. Chapter 12 describes the qualities of the leader that can create those conditions.

Puberty and Brain Development

The teenage brain undergoes huge change and development during puberty as the sex hormones flood in. Their purpose is to start to ready the brain and body for being reproductively active. A teenage girl's brain circuits to do with social interaction start to change and mature to enable her to be ready to attract a mate. As the oestrogen and progesterone kick-start the monthly cycle that makes her fertile, so she starts to experience mood swings and be particularly sensitive to her social groupings. This is a time when teenagers of both sexes begin to detach themselves from their families and find their new tribes with their peers (Damour 2016; Brizendene 2007a, p. 32; Nelson et al. 2005).

The teenage social brain is extremely sensitive to social inclusion and exclusion. In the teenage girl, in particular, social belonging becomes extremely important. Studies (McKay 2018, Chapter 5; Sebastian 2011) have shown that the teenage female brain is hypersensitive to being excluded from a grouping that is important to her. At this stage, the limbic system which, you might remember, is responsible for emotions and memory speeds up its development. The prefrontal cortex—the CEO of the brain and responsible for social cognition, judgement and analytical thought—develops more slowly, not maturing until into our twenties. This difference in growth rates between the emotional centres and the rational brain causes a mismatch which accounts for many of the ups and downs and difficulties experienced in teenage behaviour. Emotions run high driven by flooding hormones and a growing limbic system. The rational control centre in the form of the prefrontal cortex is not developed enough at this point to keep up, so there are moments of insight but often the emotional limbic system wins out. The resulting emotional roller coaster is true for both the teen themselves and others around them—as parents of teenagers will agree!

The teenage male brain, under the influence of testosterone, starts to become more aggressive and more concerned about its place in the pecking order. Not all boys want to be top dog but, on the whole, due to testosterone boys do care more about their place in the hierarchy than girls. In short, girls want to belong and boys want to know where they stand and to be the boss (Savin-Williams 1979; Moir and Jessel 1989, p. 84).

Social Context and Its Influence on Brain Development

In the first instance, our brains are designed to keep us alive and help us adapt successfully to the circumstances in which we find ourselves. The world of relationships that we inhabit as infants, toddlers and teens plays a significant role in creating our emotional and psychological patterning as we grow up. We are each born into our unique set of circumstances. Our families are situated within their particular social and cultural context and their own wider family environment. We have our place within the birth order in our families. The way we relate to our parents, siblings and wider family all play a part in how our neural patterning develops. In the fascinating, fast developing field of the new science of epigenetics (which is about how the gene expresses itself) there is growing evidence about how the life experiences of our grandparents, parents and our own experiences in our families influences how our genes become expressed.

Through applied neuroscience, it is possible for us to understand some of the key influences that individual circumstances have on the neural templating. It is a powerful tool that enables us to understand the connection between an individual's past and their leadership today. As coaches, we can help individuals to understand how their relationship patterning from early experiences supports or hinders them in becoming an effective leader. Leaders can be shown how to understand the key aspects of their own and their teams' motivational profiles. In my work with corporate clients, I have developed a tool to help them—it is called Trustprint™ and looks at how early experiences have influenced leadership identity, communication style and relational patterning.

Getting to Know the Sex of Your Brain

So, as we have seen here, it is clear that the sex of our brain is determined by three key dimensions—the structure, the connectivity, hormone levels—all of which are moulded by our experiences. These three combine to form the mosaic that is our own brain and its brain sex and it is what makes each of us the 'me' that 'I' am.

Our brains are reasonably elastic. While changing the brain is not easy it is possible and can be changed over time, but the basic sex of our brain is laid down in utero and shaped from thereon through the sex hormones at puberty and through our early experiences. The following questionnaire by Dr. Anne Moir can determine the sex of your brain. All the questions are based in underpinning neuroscientific research that indicates differences in brain sex. Find out what sex your brain is by answering the questions and scoring them (Fig. 1.2).

Test Questionnaire

It helps to understand your own learning and thinking styles if you know the balance of your own brain organisation. Simple though they are, the following questions show large differences in the brain. Answer them 'yes' or 'no' – depending on how near the answer is to your own behaviour. Inevitably these questions are generalisations, so please tick the one that most applies to you. There are no right or wrong answers – just answer quickly and intuitively.

What is your occupation? ....

	Questions (Place 'X' as appropriate)	Yes	No
1	It's easy for me to sing in tune, singing alone.		
2	When I was younger, winning was really important to me.		
3	It's easy for me to hear what people are saying in a crowded room.		
4	As a child I enjoyed going as high as possible when climbing trees.		
5	If someone interrupts what I am doing it's difficult to go back to it.		
6	I find it easy to do more than one thing at once.		
7	I find it easy to know what someone is feeling just by looking at their face.		
8	I like to collect things and sort them into categories.		
9	I solve problems more often with intuition than logic.		
10	As a child, I loved playing games where I pretended to be someone I knew or a character I had created.		
11	At school it was easy for me to write neatly.		
12	As a child, I enjoyed taking things apart to see how they work.		
13	I get bored easily so I need to keep doing new things.		
14	I don't like fast speeds, they make me nervous.		
15	I enjoy reading novels more than non-fiction.		
16	I can find my way more easily using a map rather than landmark directions.		
17	I keep in regular contact with my friends and family.		
18	As a child, I enjoyed physical sports.		
19	Imagining things in three dimensions is easy for me. For example: I can see in my mind's eye just how an architect's drawings or plans will look once built.		
20	As a child, I loved doing things like 'wheelies' on my bike.		

Fig. 1.2 Brain sex questionnaire (Adapted from 'Test questionnaire' © Dr. Anne Moir, B.Sc., Hon. D.Phil Oxon [Genetics], Human Givens Diploma. Reproduced by kind permission of Anne Moir)

If you answered 'Yes' to questions: **1, 3, 6, 7, 9, 10, 11, 14, 15, 17** score 1 point each.

('No' answers to these questions receive 0 points.)

If you answered 'No' to questions: **2, 4, 5, 8, 12, 13, 16, 18, 19, 20** score 1 point each.

('Yes' answers to these questions receive 0 points.)

Now total up your scores. Fill in your score out of 20 here:

- The higher your score out of 20, the more female your brain.
- Middle scores show a more mixed brain.
- The lower the score out of 20, the more male your brain.

Very Male   Very Female

| 1 | 2 | 3 | 4 | 5 | 6 | 7 | 8 | 9 | 10 | 11 | 12 | 13 | 14 | 15 | 16 | 17 | 18 | 19 | 20 |

Summary: Why Waste Brain Power at Work?

In our research with senior leaders across a variety of businesses we find that there is a huge amount of latent brain power inside organisations that is simply going to waste. We have been helping our clients to pinpoint where this is occurring and what to do about it.

People are generally one of the most expensive assets that a business possesses, so to be underleveraging the brain power that a business has spent a huge amount of time recruiting, training and promoting makes no commercial sense.

Why does this brain power wastage occur? Organisational culture and specific work practices (such as meetings, performance review, coaching conversations, internal communications) are often, in our experience and research, geared up to suit a more male brain better than a more female brain.

We have been measuring how much of the time the brains in the business are in a productive thriving state, with the kind of neurochemistry that produces results with the prefrontal cortex fully active. We can compare this with the amount of time the brains in the business are in a survive state, with the neurochemistry that is less conducive to performance and interference prohibiting the prefrontal cortex from fully functioning. My company-specific research is demonstrating that it is quite often the case that workplace practices and overarching culture generates thrive more of the time in those with more male brains compared with men and women more female brain sex scores. The average loss of brain potential we are finding is 30% on a working day. That is a significant amount of productivity loss for most businesses.

In summary, the three dimensions of brain structure, neural connectivity patterns and sex hormones are all a function of biological brain sex. Modern neuroscience is beginning to reveal how these differences combine with our social development to shape the individual brain that we each possess. While there is still much to uncover and understand, it is true to say that there is more than sufficient science available to us now to enable business leaders and executives to know how to intelligently access the best of differently gendered brains.

The business case for the positive impact of brain gender balance on performance is clear. Businesses that choose to understand and leverage brain gender difference will create a rich sustainable future. Harnessing brain gender diversity is the smart way to get the best of *all* the brains in the business. Knowing as much as you can about your own brain, and the brains of your team and colleagues, *is* the way to enable the conditions that create optimal energy flow through all the brains and therefore through your business.

References

Brizendene, L. (2007a). Introduction. In *The female brain* (p. 27). Transworld Publishers.

Brizendene, L. (2007b). *The female brain*. Phases of a female's life. The cast of neurohormone characters.

Cohen, S. B. (2014, February). *Neuroscience & Biobehavioral Reviews, 39*, 34–50.

Dabbs, M., & Dabbs, J. M. (2000). *Heroes, rogues, and lovers: Testosterone and behavior*. New York: McGraw-Hill. ISBN 978-0-07-135739-5.

Damour, L. (2016). *Untangled: Guiding teenage girls through the seven transitions into adulthood*. New York: Penguin Random House.

Gilmore, J. H., Lin, W., Prastawa, M. W., Looney, C. B., Vetsa, Y. S., Knickmeyer, R. C., et al. (2007). Regional gray matter growth, sexual dimorphism, and cerebral asymmetry in the neonatal brain. *Journal of Neuroscience, 27*, 1255–1260.

Ingalhalikar, M., Smith, A., Parker, D., Satterthwaite, T. D., Elliot, M. A., Ruparel, K., et al. (2014, January 14). Sex differences in the structural connectome of the human brain. *PNAS, 111*(2), 823–828. Edited by C. Gross, Princeton University, Princeton, NJ, and approved November 1, 2013. https://doi.org/10.1073/pnas.1316909110.

Joel, D., et al. (2015, December 15). Sex beyond the genitalia: The human brain mosaic. *PNAS, 112*(50), 15468–15473. Published ahead of print on November 30. https://doi.org/10.1073/pnas.1509654112.

Mac Giolla, E., & Kajonius, P. J. (2018, September 11). Sex differences in personality are larger in gender equal countries: Replicating and extending a surprising finding. *International Journal of Psychology*. https://doi.org/10.1002/ijop.12529.

Mckay, S. (2018). *Demystifying the female brain: A neuroscientist explores health, hormones and happiness* (pp. 633). London: Hachette UK.

Moir, A., & Jessel, D. (1989). *Brain sex: The real difference between men and women* (p. 50). London: Mandarin.

Nelson, E. E., Leibenlfut, E., et al. (2005). The social reorientation of adolescence: A neuroscience perspective on the process and its relation to psychopathology. *Psychological Medicine, 35*(2), 163–174.

Savin-Williams, R. (1979). Dominance hierarchies in groups of early adolescents. In D. McGuinness (Ed.), *Dominance aggression and war* (pp. 131–173). New York: Paragon House.

Schore, A. N. (2000, April). *Attachment & Human Development, 2*(1), 23–47.

Schore, A. N. (2001). *Infant Mental Health Journal, 22*(1–2), 7–66. Michigan Association for Infant Mental Health.

Sebastian, C. L., et al. (2011). Developmental influences on the neural bases of responses to social rejection: Implications of social neuroscience for education. *Neuroimage, 57*(3), 686–694.

Southern, A. L., Gordon, G. G., Tochimoto, S., Pinzon, G., Lane, D. R., & Stypulkowski, W. (1967, May). Mean plasma concentration, metabolic clearance and basal plasma production rates of testosterone in normal young men and women using a constant infusion procedure: Effect of time of day and plasma concentration on the metabolic clearance rate of testosterone. *The Journal of Clinical Endocrinology and Metabolism, 27*(5), 686–694. https://doi.org/10.1210/jcem-27-5-686.pmid6025472.

Southren, A. L., Tochimoto, S., Carmody, N. C., & Isurugi, K. (1965, November). Plasma production rates of testosterone in normal adult men and women and in patients with the syndrome of feminizing testes. *The Journal of Clinical Endocrinology and Metabolism, 25*(11): 1441–1450. https://doi.org/10.1210/jcem-25-11-1441.pmid5843701.

Torjesen, P. A., & Sandnes, L. (2004, March). Serum testosterone in women as measured by an automated immunoassay and a RIA. *Clinical Chemistry, 50*(3), 678. Author reply 678–679. https://doi.org/10.1373/clinchem.2003.027565. pmid14981046.

US National Library of Medicine, National Institutes of Health, Chiron, C., Jambaque, I., Nabbout, R., Lounes, R., Syrota, A., & Dulac, O. (1997, June). The right brain hemisphere dominant in human infants. *Brain, 120*(6), 1057–1065.

Willerman, L., Schultz, R., Rutledge, J. N., & Bigler, E. D. (1991). In vivo brain size and intelligence. *Intelligence, 15,* 223–228.

Witelson, S. F., Beresh, H., & Kigar, D. L. (2006, February). Intelligence and brain size in 100 postmortem brains: Sex, lateralization and age factors. *Brain, 129,* 386–398.

2

Conditions for Optimal Brain Function

Paul says:

The Strong Case for Biology as the Basis of Behaviour

Towards the end of 2014 Dick Swaab, a highly distinguished and decorated neurobiologist and doctor in the Netherlands, published *We Are Our Brains: From the Womb to Alzheimer's*. 'A fun, wild ride through the big science of the moment', said the *Sunday Times*, being very careful not to commit itself to any opinion as to whether it agreed or not.

It soon became a best-seller, but Swaab had to stand firm against all the social pressures that came his way, including death threats, for stating clearly and unambiguously how it is that each of us becomes who we each are *because* of our biology. Are we each who we are through experience?—the social science's essential view. Or are we each who we are because our biology says so? Swaab is uncompromising in the views that he arrived at through his research findings from the 1970s onwards. His Wikipedia entry records his now often-quoted aphorism: '… our brains are not things we have, but rather brains are what we are'.[1]

---

[1]https://en.wikipedia.org/wiki/Dick_Swaab.

© The Author(s) 2020
K. Lanz and P. Brown, *All the Brains in the Business*, The Neuroscience of Business,
https://doi.org/10.1007/978-3-030-22153-9_2

A Story to Start a Metaphor May Help

A good number of years ago, living in an old farmhouse on the Somerset levels, we were lucky to find Dave, a retired brick-maker, to odd-job around the place, and his wife Eileen to keep general order. Dave loved nothing more than to stop and talk about the way things used to be. He had been a teenager throughout the war, living on the Somerset coast, and had spent the whole of his working life in a brickyard.

The clays in that part of the world are a blue-grey lias. Dave still owned the spade that he had last used in the brickyard. It was of the same pattern that he had first used as a fifteen-year-old when he went there as the war ended. Less wide than a normal garden spade, and a good deal flatter, but with a pronounced bend in the willow shaft just above the metal socket where wooden shaft and metal blade were bonded together, Dave showed me how he could make four separate sharp thrusts of his spade into clay and out would come a perfectly formed brick shape ready for hardening in the brick ovens. He had done that all his working life. The bend in the willow shaft added power to the leveraging out of a brick from its clay stickiness.

Dave's bricks, and the pile he must have cut out of the clay in fifty years of a working life, created this metaphor for the brain.

Imagine a huge pile of bricks. At the same time, somewhere in a small unused space, hold in mind that each one of the huge pile of bricks is a brain cell: one of the 86 billion neurons that comprise the intact adult brain, each one a miracle in making pathways of electro-chemical circuits. Now go back to the bricks.

From the pile of bricks it would be possible to construct what look like from the outside a row of small, mean, cramped standardised cottages or a single gracious and spacious Georgian mansion.

Imagine going into one of the cottages. To your surprise the welcome is warm, it is beautifully and tastefully decorated and there are small treasured but carefully bought items that make a collection of silver connected to the old wine glasses that the owner has carefully found. They reflect the twinkle in her eye. The delight of having been there lingers with you long after your visit is over.

Imagine now going into the Georgian mansion. To your surprise and slight terror you find it is cold and dark and dirty. There seems to be no-one in. You explore, hesitantly. Some doors are locked. Others swing open on to untidy rooms. The main corridor disappears in the gloom and the stairs look less than inviting. The memory and surprise of that occasion also linger long after you have been glad to leave.

Think now of each house as a person. Perhaps you can make it a real person you know or have known.

As with the bricks, human beings come into the world with a great **supply** of brain cells ready to be put to use but the essential architecture has already been created during the second two trimesters of life. The neurochemicals that are triggered as part of the extraordinary developmental process that changes an embryo into a foetus and then into a baby pushing a way into the world have defined the basic architecture of the brain, as have the neurochemicals and hormones coursing through the mother's blood and the baby's brain. Experience creates the person's interior, but biology creates the structure within which that interior can be developed.

Why does it matter, though, whether it is biology or experience that shapes us when it seems to be both? Are we stuck with an either/or approach—that our behaviour is determined biologically or, fundamentally, socially? Or is that only for academics who can't move on?

The answer is that if we want to get to *explanations* for human behaviour then science in the form of neurobiology and the new sciences that are emerging from that discipline are the only way we know of getting there. If we are happy enough with *descriptions*, then any amount of social theorising and observation will do. The West—and now most of the rest of the developed world—has majored on replicable knowledge established through experimental science as the way of advancing and agreeing what it knows. This is what is happening this century in completely re-defining our understandings of human behaviour. The essential 'me' has been defined before 'I' came into the world. Once there, becoming **identifiably** 'me' is about how we create an interior.

A Newtonian Moment in Time

There are times in history when patterns of thought change radically as new understandings of the world develop. We are in the early stages of such a shift. A historical comparison might make this clear—that we are at an extraordinary point in time in the early development of knowledge about the brain and what underpins human behaviour. Understanding the brain is moving from *imagining* how it works to *knowing* how it works.

That's a big shift. It starts about five hundred years ago.

In 1508 Michelangelo started what was to become four years' work, creating the religious intensity that is the ceiling of the Sistine Chapel that had itself been finished only thirty years before. His God with outstretched arm touching Adam's outstretched arm across the abyss that defines the void of Creation has become one of the great images of High Renaissance art.

The ceiling portrays many aspects of a divine authority, of which the Pope was the earthly representative. But that was being called into question. Neither Michelangelo nor Pope Julius II who commissioned him could have known that their world was on the cusp of a profound change. The old religious verities were to start giving way to new scientific truths, though the process would take a hundred and fifty years.

Ordained to the priesthood the year before Michelangelo started painting his Sistine Chapel ceiling, five years after it was finished Martin Luther in Germany was questioning many of the mercenary and other practices of the Catholic Church. We now know that the beginnings of what became Protestantism had occurred, which in turn powered radical processes of new thought. The Pope's demand that Luther renounces his own thoughts in favour of papal authority served paradoxically only to unleash a new freedom. Henry VIII's break with Rome and the beginnings of what became the Church of England, occasioned by the un-granted wish for the Pope to pronounce a divorce, was part of the same questioning of papal authority through that sixteenth century.

From classical times until the Renaissance the way of resolving arguments and arriving at an agreed truth had been (if not by force) through disputation. The skill of disputing was taught as logic and rhetoric backed by philosophy. It's what the universities of the time were about. But with the gradual erosion of papal authority a new kind of intellectual freedom started appearing that looked for verifiable truths. When, in the early seventeenth century, the papacy failed to silence Galileo who argued that the bible was an authority on faith and morals, not on science, it was becoming apparent that what had been philosophy was starting to become the experimental philosophy that became physics: that understanding the universe was not through divine revelation but by working out and agreeing what the laws were that could *explain* what was happening.

It was mathematics that powered this shift, with its apotheosis at the time in Isaac Newton's *Principia Mathematica* of 1687, a quarter of a century after the founding of the first scientific society in 1660 as the Royal Society in London.

And now mathematics is powering a new shift. Allied to biology and harnessing the information sciences of the twentieth century, it is helping create an algorithmic understanding of the brain and human behaviour that is a radical shift from the disputatious twentieth-century understandings.

Why disputatious?

Well, looking back over the psychologies of the twentieth century, it turns out there were more than fifty identifiably different 'schools' of psychology. Psychologists did not agree on anything that identified them individually or collectively as scientists or, in the various applied psychology professions (clinical, educational, forensic, counselling, coaching, sports, health and so on) as having a common, shared background. From a plethora of theories a very large number of different therapies emerged, but based on no common, scientifically-validated frame of reference.

At the beginning of the twentieth century, Freud had drawn attention to the fact that early life experience goes on reverberating throughout life. He re-formulated something that had long been known, and that found one expression in a poem of Wordsworth's in 1802, where he used the phrase 'The child is father of the man' a century before Freud's main writings started.

In his quest to create an understanding of how the mind works, what unfortunately Freud elaborated were a set of theories, based substantially on imaginings about the repression of sexual desires, that are now known to be completely false. A recent book by Frederick Crews called *Freud: The Making of an Illusion* details a forensic pursuit of finding out what was fact and what was false in Freud's writings about his clinical observations. The evidence of falsity is overwhelming. Freud wrote about what he believed *ought* to be happening in therapy, not what was *actually* happening. But in the absence of any other such prolific writings on the subject of his interest, Freud's theorising established itself in common discourse throughout the twentieth century as if it were fact.

Why is all this important? It is because Freud created and represents an ideology, a set of belief systems, for understanding human behaviour. We can now see, with just a little hindsight, that psychologists attaching to one of the many theoretical systems that developed after Freud were in fact creating a set of psycho-theologies. What they offered in therapy or taught in universities was what they believed, not what was verifiably true. But now there is a clear shift to scientific systems in our understanding of the brain and human behaviour. Without wishing to stretch the analogy too far, it is as if Freud was a Pope and Swaab a Galileo: one relying on dogma and ideology, the other risking the slings and arrows that scientific understandings may create.

Not that, in writing this book, we fear slings and arrows. The temper of the times has become too interested in the brain and how it works for there to be serious controversy about the evidence that underpins our thinking.

Where there may be honest differences, though, are in the conclusions that we draw about the future of women in organisations and what women might do for organisations when some glimmer of light illuminates male blindness as to what the differences are that women bring to organisations. For they bring feminine energy which, we shall argue, might create very different approaches and organisations from the ways men go about things.

***

Brains Emerging into Being Observable Organs

What exactly *is* this thing that we call 'a brain', though, and that seems to have such power? Taking it just for granted won't do, any longer, now that we can actually *see* it working.

1980 marks the point at which it became possible to create an image of the live brain. Although for nearly fifty years before that it had been possible to look at tracings of electrical activity in the brain—EEG's or electro-encephalograms—the only way to *see* the brain had been post-mortem or during surgery. Neither of those are very good circumstances for systematic research into what the brain actually *does*. Simply seeing the whole brain working inside its protective skull was not, until 1980, an option; and in any event looking at the brain post-mortem or during surgery is not vouchsafed to many people and of little immediate interest to the brain's owner.

It was at Aberdeen University that the first MRI (magnetic resonance imaging) machine to create clinically useful black and white images was created. In less than forty years the technology for looking at the detail of the brain has developed at an extraordinary rate. Now brain activity right down to single brain cells can be seen, and in extraordinary colour. The big overriding conclusion at present is that the more that is known the clearer it becomes how little is known, even though it is thought that in the last twenty years more has become known about the brain than in the whole of previous history.

Take the question of how is information transmitted around the brain, for instance.

Since the late 1940s, the working model of how the brain functions is that there are an enormous number of pathways making circuits: that the pathways are formed by experience triggering electrical potentials that release chemicals that transport information; and that the more often any circuit is triggered the stronger it becomes. This model suggests that the brain is like an immensely complex road system with major roads and minor byways, major junctions and perhaps some lay-bys and maybe with little lanes blocked up with weed.

But recent research suggests a completely alternative model. Think of brain cells being like the grains of sand on a beach. A small wavelet (a little sound) or a huge wave (suddenly seeing the person you instantly fall in love with) rearranges all the grains but then stores that pattern so it becomes part of the you that experience continuously re-creates and reinforces. Or maybe both models are right, and co-exist in ways that are far from known as yet. If an enormous amount about the detail of how the brain works not yet clear, the directions of travel in the worldwide quest for understanding how nearly eighty-six billion brain cells weighing in at nearly four pounds and powered by electricity and chemistry works *have* now become very clear. And perhaps the most significant is to make it clear that we human beings are, like the rest of the physical universe, primarily *energy* systems and not, as the twentieth century supposed, psychological systems. It is that fact—that we are energy systems, a completely different perspective than that which the twentieth century created about the human condition—that makes it possible to think, talk and write intelligently and from a scientific perspective about the differences between men and women.

That's the remarkable shift in thinking about humans that is going on.

As we contribute to that thinking bear with us for using the terms, 'men' and 'women', 'male' and 'female', 'feminine' and 'masculine'. Despite there now being at least fifty-three designations for gender variations, the common frame of reference that has done the evolution of human beings pretty well for at least two million years, and the rest of the mammalian world too, as well as the reptilian world for much longer, is the frame of reference we adopt here.

If, at the arrival of baby into the world, prior scanning (in the developed world, and then only for some) has not already established its sex, the two questions of 'Is it alright?' and 'Is it a boy or a girl' are those most parents ask immediately. And though we now know that apparent genital sex is not a completely reliable guide to gender, in the vast majority of cases it is though, as is fundamental to this book, what the biology of uterine development may have been up to in sexing the brain and sexing the body is part of an increasing understanding that the whole business of defining who we each are is more complicated than once seemed to be the case. The biblical account of creation in Genesis that says 'Male and female created He them' simply did not know about the dimensions and variations that might become known, though they have probably existed since the beginnings of humanoid ancestry.

So having borne with us, let us return to the difference between the twentieth and twenty-first centuries with regard to understanding people. For that is what makes it possible to begin to think about the differences between men and women in their corporate roles.

***

Two New Brain-Related Sciences to Take on Board

It will be useful to take on board at this stage some information about two of the new sciences that have developed around the neurosciences. The first is connectomics, which won't detain us too long. The second is epigenetics, which becomes fundamental to our understanding of the differences between men and women.

Human beings have always been able to look at the body and its various parts. Different cultures and religious systems have managed, permitted or denied the use of that capacity in widely differing ways. But as the opening of curiosity about the natural world developed, representing the natural world accurately became a greatly appreciated skill. Leonardo da Vinci was a master of it and, to set him in historical context, he died in 1519, within the decade after Michelangelo had completed the Sistine Chapel ceiling and just after Luther had challenged papal authority.

Leonardo da Vinci's drawings of the brain are wonderfully detailed and accurate. They represent the early stage of modern neuroscience, which asked the structural question 'What does it look like?': which in turn begged the question 'And how shall we name the parts?' The specifics of the brain have acquired a remarkable nomenclature.[2]

The second stage is often given a start date of 1848. Phineas Gage, a twenty-four-year-old foreman in charge of blasting rocks during railway construction in America's Vermont, was using a long metal rod to tamp down explosive powder in a cavity prior to igniting a fuse that would cause an explosion to shatter the rock. Unfortunately the metal rod inadvertently created a spark against the rock that caused the powder blast to occur and the metal rod to be fired with great force through Gage's left cheek behind the eye and then out through the top of his skull, the bar landing some

---

[2]https://en.wikipedia.org/wiki/List_of_regions_in_the_human_brain.

eighty feet away. Gage remained conscious, sat in the back of a cart to be taken to the nearest small town, where he told an attending but disbelieving physician what had happened.[3]

Gage survived, but with substantial personality changes. It is reported that, when he tried to return to work, his fellow workers said 'Gage isn't Gage any more'. From having been a model and efficient and much-valued worker he became the opposite, started drinking heavily and lost his marriage. He eventually went to Chile to drive a stage-coach, but returned to the USA when he started having fits and in due course died, having survived twelve years after the accident.

Survival after a penetrating head wound of such severity was very rare at the time, and medical attention focused on Gage's temperamental and personality changes. So a second stage of interest in function starts. 'What does each bit do?'

When the power of computing in the twenty-first century got coupled to increased capacity to image the brain, however, a new science developed asking the question: 'But how is it all connected?' That is connectomics. It produces the most stunning images, of which Sebastian Seung, originally at MIT and now at Princeton, is a particular early and prolific innovator. His book of 2012, *Connectome: How the Brain's Wiring Makes Us Who We Are* is a classic of its kind. His lab also created the most remarkable internet crowd-source game, called eyewire (https://eyewire.org/explore), through which thousands of individuals have contributed to the science of unravelling the pathways of the brain.

If you stop and think about your brain working every millisecond of the day and night for the whole of your life, just like your heart though without being able to hear its rhythms in the same way, putting together or sorting out all the data that pours in all the time as well as finding what is stored there when it needed, and that is does that seamlessly; doesn't that seem a remarkable fact? Despite the brain having many specialised areas of function—sight, for instance, or hearing, but also thousands more—it sends out signals as if there were no boundaries and no conflicts across the boundaries.

Think of applying that fact to organisations. A moment's reflection about the sheer waste of energy that happens in inter-departmental disputes within so many organisations might create a longer pause for thought, asking how could an organisation operate seamlessly too? How could its energies be used not for fighting and interpersonal competition but for much

---

[3]https://en.wikipedia.org/wiki/Phineas_Gage.

more satisfying productivity? That's the importance of connectomics to the future of organisations. And if, as we go on saying, this book is about the differences between the way men and women might use their energies, then the new science of connectomics might be part of the working model for the twenty-first-century organisation that turns feminine energy into better account than simply striving to be the equal of a man. Connectomics underpins co-equality. It's how we do things *together* in the best possible way and the evidence of it working well is e-quality.

The other new science is called epigenetics. There are two branches to it. The first, which will not detain us because it is entirely based in the laboratory, is called molecular epigenetics. The second we have called, in order to distinguish it for our purposes here, perceptual or social epigenetics. It goes like this.

In 1953 Watson and Crick gave us the early stage image of the double helix as a way of being able to visualise the gene and how it worked, based in no small part on the pioneering work in crystallography of Rosalind Franklin at King's College, London. The second half of the twentieth century became suffused with the science of the gene. In popular imagination it became the controller of every aspect of behaviour, and huge prospects were held out for understanding the gene as being the route to curing many intractable illnesses.

As is often the case with science, early understandings give way to understanding only that there is more understanding to be done. And part of that, with regard to the gene, has been the understanding that the gene can express itself in a huge variety of ways.

What does that actually mean? Here's a thought that we especially like from Nessa Carey's book *The Epigenetics Revolution*. Bear with us again while we get through a little explanation to the heart of the matter.

Why We Are Each Unique

In every cell in our body we have DNA that carries the information as to how, in every element of its being, the body is to function. It does that from twenty-four chromosomes, but those chromosomes can be expressed in millions of combinations. So take as a comparison the English alphabet with its twenty-six letters. Every single book that was ever written in that language has re-arranged the letters to convey meaning that is different from every other book.

It looks as if our bodies encode experience through the way the genes express themselves in the giving of instructions as to which proteins are to be bound to which bit of experience. The consequence of this is that once we have had an experience there is a pattern for it to be repeated: the more it is repeated the more the pattern becomes established and the resulting behaviour, whatever it is, becomes a part of 'me': and as a pattern it is laid down in the neurochemistry of the brain and perhaps the chemistry of the whole of the body.

Speculatively, it may be that the basic emotions with which we come into the world as the basis of making sense of that world are the simpler forms of this epigenetic process. Then the gradual mixing of the emotions into feelings that, with their associated neurochemistry, form the basis of perception gives meaning to individual experience.

The essential point, however, is that once experience and neurochemistry are bound together *then* they dictate, organise, determine our *individual* behaviours. And it is that that makes us unique. And if, as is undeniably the case, women have different mixes of chemicals, neurochemicals, hormones and neurohormones than do men, then their behaviours will be different and their energies expressed differently.

Bruce Lipton has been an early pioneer in this field. And though perhaps a little messianic for some tastes in his mixing of science and popular self-help these days, his understanding of the way perception initiates the neurobiology of action is profound.

Take a simple example. Imagine that as you are leaving a busy street to walk into your favourite store, coming out of the store is a very old friend who, in the urgencies of life, you happen not to have seen for two or three years. The immediate reactions of both of you on seeing each other simultaneously happen without a moment's conscious thought. Yet they happen. That is due to the neurochemistry of perception.

Imagine, alternatively, that as you are about to walk into the store you see whoever the person is who, for whatever reason, is the very last person you would wish to meet ever again. Every bit of sensation and overt behaviour you have at that moment would be quite different from the way you greeted your old friend. No instant recourse to a coffee shop for a long and unexpected catch-up on this occasion, for sure.

And just to finish off this understanding of the neurobiology of perception, bring into your imagining of both the above occasions the fact that another friend was with you but who knew neither of these people that you met by surprise, nor their significance in your life. Your friend's reaction to either of the people would not be the same as yours. S/he would of course, see your reactions and might empathetically enjoy the first and be worried for you by the second. But the people who triggered the reactions in you were not in any way of the same significance to your friend. So the reactions would not be the same.

So what?, you might say. What you are describing is just a perfectly ordinary set of events that, in different contexts, are the stuff of everyday life. And we have known about them for *as long as humans have had interactions with each other.*

That's true, of course. But what the science of epigenetics is now telling us is that *perception arising from experience* is what sets up the neurobiology of behaviour. So what infancy, childhood and adolescence are about is priming the system neurobiologically to make each individual the individual that s/he is to become. That neurobiology is what earlier, when exploring the brick metaphor, we meant by internal decoration coming from experience. The experience is encoded neurochemically, as if we were painting a room. When it's finished, that's the way it is. It might have been many other ways, but it isn't. What we think of as social becomes profoundly biological.

This, then, is a remarkable strength and remarkable limitation of who we are. As adults we need to be who we are because otherwise we cannot function. And yet, built into the biological design for all life, is the possibility of also being adaptive. Otherwise we would not be here at all.

Which takes us into thinking about surviving, thriving and what flow is all about.

References

Carey, Nessa. (2011). *The epigenetics revolution: How modern biology is rewriting our understanding of genetics, disease and inheritance.* London: Icon Books Ltd.

Crews, F. (2017). *Freud: The making of an illusion.* London: Profile Press Ltd.

Seung, S. (2012). *Connectome: How the brain's wiring makes us who we are.* Boston, MA: Houghton Mifflin Harcourt.

Swaab, D. (2014). *We are our brains: From the womb to Alzheimer's.* London: Allen Lane/Penguin.

3

Survive, Thrive and Flow

Paul says:

At lunch in Dubai, Sarah was laughing at the memory of being eighteen. It was 1990, her first week at university, and she was one of two-hundred and seventy freshers starting out to become engineers of one kind or another. The first year was a general year for them all. Specialisation started in the second year. She already knew she wanted to become an electrical engineer.

What she didn't know was that, nearly thirty years later, she would be CEO of a heavy electrical engineering company doing contract work in some of the most inhospitable environments where oil and gas are found. Frequent trips to frozen northern Russian and the frontiers environment of Kazhakstan were part of ordinary everyday life; to be managed alongside a husband, a home, a daughter aged four and twin boys just a year old. As she had said earlier, when talking wryly of the size of job she had and a commitment to growing the company in two years to twice its current size of a quarter of a billion dollars revenue, what an advantage it was being a woman because she could enjoy her job and worry about whether the fridge was full at the same time.

But that wasn't the immediate cause of the laughter. The conversation we were having was about what the differences are between male and female ways of getting through the world. What Sarah was recalling was that, of the two-hundred and seventy freshers, exactly fifty of them were girls: and in the first week of term the girls only had been invited to a special lecture by the Dean of the engineering faculty.

© The Author(s) 2020
K. Lanz and P. Brown, *All the Brains in the Business*, The Neuroscience of Business, https://doi.org/10.1007/978-3-030-22153-9_3

The lecture turned out to be an hour's impassioned plea by the Dean that they all re-consider their wish to pursue engineering studies. It was, he made plain, not the province of girls to go into engineering as a profession, so what was the use of studying it? Taking no account of the fact that every single individual in the group had gained school leaving certificate high marks in at least two science subjects and higher maths as the basis of being there at all, he suggested that their future might be much better if they went across to the arts or social sciences faculties in one form or another before it was too late to make the change.

'What did you all do?' I asked.

'Completely ignored him', she said. 'We laughed about it as soon as he left the room. We all knew what we had chosen and for each of us why we had chosen it. Not a single girl moved to another faculty, and forty-seven of us finished the next four years which was a better average than for the boys'.

'It's inconceivable isn't it', she went on to say, 'that such a thing could happen now? It's not quite thirty years ago. I wonder what it would have been like thirty years before that, in 1960? Maybe we were lucky and right in the middle of a shift of being able to be who we wanted to be. We didn't have to believe what we were being told just because some man in authority was saying it. But I think it would have been much harder a generation before. We've come a long way'.

'Far enough?' I said.

'Come off it!' she said. 'You're not going to get me going on that one. That's a typical man's question. I struggle with that kind of question all the time in this organisation. Men want answers in a binary way: "yes" or "no", "this" or "that", all the time. Of course on technical matters we deliver answers to our customers that are better right than wrong. We do that all the time. But at Board level it's different. Getting a conversation going there that's not just about solving problems but is having time to think, looking for solutions, trying out ideas without immediately wanting answers, is one of the hardest things in my day'.

That took us off into a quite different kind of discussion, and it started with the nature of identity. Coaching lunches are a great place for not finding answers but edging towards insights.

***

The Search for Identity as the Basis of What the Differences Are

In the middle of 2018, I was in a professional development session and had started a discussion about the differences between men and women. A member of the group made the challenge that is always the most difficult when it comes up in such discussions. 'What *exactly* do you mean when you talk about the differences between men and women?' she said.

To my own surprise I heard myself saying something that had never quite formulated itself before. 'This is the real difference', I said. 'Men do what they do in order to find out who they are. Women do what they do because of who they are'.

We establish our individuality and identity in different ways.

A pin could have been heard dropping in the reflective quiet that immediately pervaded the whole of the room. It felt, and goes on feeling, as if there is an essential truth in that sudden aphorism. As a truth it's what the future of organisations is premised on—finding the ways of working that take the differences and best of both.

But that's the big picture. We need to get into some detail first to find out how each person's system would work to create that truth.

We established in Chapter 2 that we human beings are fundamentally energy systems, not psychological systems. Complex and adaptive though we are, a practical question is: What is the source of that energy and how is it managed and used?

The brain relies on blood transporting oxygen and glucose around it as the physical sources of supply that give the brain its energy. The body as a whole runs off about 95 watts of energy. The brain takes up 25 watts of that. So though the brain is only 4% of body mass it consumes more than a quarter of the available energy. It's a seriously energy-hungry organ.

In thinking how little we know about how to manage that energy, though, and as a small aside, think of the way gyms and sports centres have become a part of so many people's lives over the past three or four decades. Deliberately looking after one's body has become a normal, wide-ranging social activity in a way that five decades ago was not on the agenda at all. Sports like tennis or cricket or athletics were seen at that time as sports, to which health was an incidental benefit but not a primary motivation. One of the consequences of this big shift in appreciating the value of being physically tuned has been an increasing appreciation of the detail of the way the body works.

So far as setting out to enhance brain performance is concerned, the evidence for its effectiveness is equivocal though the market for it is there. Dr. Tokuhama-Espinosa (2019) has recently reported that the market in 'Brain training' grew from $600 million in annual revenues in 2009 to more than a $1 billion in 2012 and is projected to reach $4–$10 billion by 2020, according to a 2016 promotional video by SharpBrains, a brain training enterprise.

That prediction for 2020 seems to have a wide margin of error in it for the short time-slot ahead when it was made. And it's a little way behind the Wellness Creative Co's late 2017 predictions for the physical fitness industry's expected global value of $87.5 billion in 2018. The safest conclusion at present is that diet and sleep are the two best ways of maintaining a healthy brain.

Sleep is crucial because an extraordinary cleansing process happens during sleep. The toxins and left-over, unused chemicals and other waste matter from 86 billion brain cells get flushed out overnight by the cerebro-spinal fluid. But it only happens in the deeper phases of sleep and is part of what makes deep sleep feel so good on waking. It used to be thought that the fluid that surrounds the brain inside the skull was simply a clever means of lessening the adverse consequences of any impact damage that the brain might have from banging against the hardness of the skull. Now that fluid is known to have a much more active process. Poor sleep—sleep that does not leave one feeling refreshed on waking—leaves toxins and rubbish active in the brain. It's like having a kitchen that is not properly cleaned and where things aren't put away properly. Everything's there but it works sub-optimally.

And diet is vital because one of the major discoveries of the twenty-first century is just how much interplay there is between the gut and the brain.

A new science called gut neuro-microbiotics has developed in order to start to understand how the way that trillions of minute organisms to which the gut plays host actually influence the way the brain works.[1] A typical experiment to demonstrate this takes a rat from a strain of rats that have been bred to be anxious, cleans out its gut completely and then re-fills the gut with the gut flora and fauna from a rat that has been bred to be placid. The previously anxious rat becomes much more like those of the placid strain. Trials are being started to see if such findings can be useful in the treatment of some human conditions, especially those with apparent inflammatory origins like rheumatoid arthritis. A recent book by the Professor Psychiatry at Cambridge, Dr. Edward Bullmore (2018), is especially good in describing how inflammation invades both the body and the brain.

---

[1] See https://en.wikipedia.org/wiki/Human_microbiota.

And Now It Gets Emotional

But all of that, though about specifics in some ways, doesn't quite get us close enough to the central issue about how to create flow. For that we have to get some understanding of the emotions, because they are what drive behaviour.

Unfortunately, the scientific understanding of the emotions is a mess. This is because the vast majority of the literature about the emotions has come from the social sciences rather than the experimental sciences. In order to start to sort out that mess it has been proposed that there should be at least a working agreement about what it is scientists are talking about when they explore whatever it is 'the emotions' are. Medicine proceeds on the basis of scientific discoveries. The law has its assumptions continuously tested and refined by the courts. Accountants agree and make public what their operating standards are as the basis for shared professional understandings. But the social sciences seem to be able to exist without such disciplines. They function as descriptive rather than explanatory bodies of knowledge, which alas seriously limits their utility value and why psychology has made so little impact on the world in the last hundred-and-thirty years of its existence compared with, say, engineering over the same period of time. Psychology hasn't done the equivalent of getting us to the moon: and arguably has not even, metaphorically, decided whether or not a moon that one might be trying to reach exists at all.

So a process was started in early 2018 to try to start to remedy this situation. A working model of the eight basic emotions that had been in slow development for the previous ten years[2] was published and named as *The London Protocol of the Emotions*; with the proposition that it would at least be a statement with a strong scientific base to it that might, in due course, become a common starting point for a comparative literature. Figure 3.1 shows it as printed.[3]

If the first purpose of the London Protocol to suggest a way of stabilising research on the emotions, its almost equally important second purpose is to

---

[2]In particular, Carrie Coombs, Jane Upton and Tara Fennessy, all CPD students at one time or another and equally colleagues, clarified some points of display or content in that process.

[3]A literature review that started with a search limitation of the decade 2008–2018, though got somewhat larger as back references were also pursued, and that had the aim of seeing what the literature said about emotions, feelings, moods and states was stopped after fifty-seven pages of references had been accumulated, with a sense that there was enough data from which to begin to draw some assumptions and make some proposals (Brown and Dzendrowskyj 2018).

8 BASIC EMO+IONS

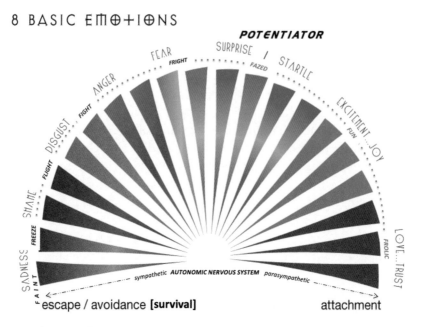

Fig. 3.1 The London protocol of the emotions (*Source* https://iedp.cld.bz/Developing-Leaders-issue-29-Spring-20181/26/, © Paul Brown 2018)

make a popular, easily accessible but scientifically valid frame of reference available for general use. Being able to name one's emotions, bring them into conscious awareness with a good deal of precision, and regulate them to get maximum value out of them through such a framework of understanding, is that purpose.

Embedded in the word 'emotion' is its hyphenated form, *e-motion*. That slight perceptual shift tells us that embedded in the word 'emotion' has forever been the concept of e-motion: energy-into-action. It is clearly time that we stopped considering emotion as being in opposition to rationality. Both are part of, and drive, a whole complex system called 'me'.

It was Antonio Damasio (2000), one of the late twentieth century's perceptive prophets of the coming neuroscientific revolution, who made the brilliantly simple suggestion that we start to think of the basic emotions like the primary colours. From three primary colours—red, blue and yellow—the whole of the colour palette comes. Similarly, mixes of the eight basic emotions create the whole of the feeling system.[4]

---

[4]They also create moods and states. Emotions, feelings, moods and states are words that get used interchangeably. That does not make for scientific precision or understanding.

We come into the world with the potential for any one (or more) of these eight basic emotions to be triggered by whatever external or internal circumstances exist; and to which the brain at whatever stage of development it is makes the best adaptation that it can. That is the primary job of the brain. To make the best adaptation that it can.

Humans take a long time to develop the use of all the equipment that evolution offers them. Simple reflexes—coughing, sneezing, elimination, going to sleep, waking up—show that the baby is alive and functioning. It's the same with a puppy. There's a reflex yelp if its tail is accidentally trodden on and a baby cries with pain and a reflex startle if a hand or lip comes into contact with something much too hot. That's the most basic survival stuff. Without it we would not be here at all.

What seems to be very special about human beings is the very long period of creating relationships that lies ahead of the day-old baby. And for that evolution[5] has provided not only survival emotions but attachment emotions also—the two dimensions on the right-hand side of the fan, excitement-joy and trust-love. Without the anticipation that startle/surprise produces and the escape/avoidance of fear, anger, disgust, shame and sadness (fadss if you like mnemonics) there would be no way of understanding the greatest of human experiences that the attachment emotions of excitement-joy and trust-love create.

But the attachment emotions have no primary survival value. It's why they can so easily be over-ridden by the survival emotions. And that's why also the most simplistic theories of motivation—carrot and stick—never make for real success. The carrot keeps the beast alive waiting only for the next stick. The brain remembers the stick, in whatever corporate form it comes, infinitely more than the carrot; for the brain is on the alert all the time for external events that would trigger a rush of survival activity.

In a baby it is the emotions attached to consistent experience that shape the pathways of the brain and, through experience attached to heredity, makes that baby the individual s/he becomes. That is why we are all different, just as every painting is different and every published book is different.

The same materials create amazing differences. Twenty-six letters of the alphabet can create a book that is pornographic, the most romantic novel,

---

[5]Whenever we invoke 'evolution', it's worth remembering that none of us were there when it was happening. All statements about evolution are necessarily inferential. Evolutionary biologists can get rather more inferential than even psychoanalysts' imagining of what was supposed to have happened in the past.

the plays of Shakespeare, or a scientific text impenetrable to anyone but a specialist in the same subject. Similarly, the eight basic emotions can create a brain whose only language is full of expletives and filled with escape-avoidance emotions; or, depending on how they are triggered, can create a brain that searches for the greatest possible fulfilment and expression of being human throughout the whole of its adult life. The materials used by both brains are exactly the same. The way those materials are put into use and create the person is different.

And this brings us to the crux of the difference between male and female. The materials used to create male and female are exactly the same. The way they are put to use is different. And it is in the aphorism that describes the search for identity—becoming *Me*—that the difference lies.

'We establish our individuality and identity in different ways', was the statement that followed on from the aphorism that appeared, apparently un-thought but obviously long-gestated. *Men do what they do in order to find out who they are. Women do what they do because of who they are.* One major consequence of that is that the way energy gets displayed is different. That is what is at the heart of the social perception of male-ness and female-ness, masculinity and femininity. And if, as is also basic to this book, the way energy is used or wasted is what creates profit and outcomes in organisations, then it follows that *if*, as we are proposing, male and female energy is different, the experience of the way energy flows would be different.

Let us briefly define what we mean by energy in this context. Human beings coming to the workplace each day for many hours, bring in their capacity to feel motivated, focus their attention and problem-solving brain power on the tasks in hand and to deploy these to get things done at work. At the simplest level they will either do this with a spring in their step and a smile on their face because they feel excited at the prospect (a high energy state) or they won't (a low energy state). A brain that is being forced to operate out of its own individual natural thrive state, will not be in flow. To a greater or lesser extent it will be in survive (not fully accessing the prefrontal cortex) and there will be less attention, focus and motivation towards task completion. Being in such a low energy state is rather like driving a car with the handbrake on.

It's not that women cannot be *just* as interested as men in profit or other outcomes. It's that instead of wasting their corporate energies being the best men they can be, why not turn those energies into being the real women they can be and see how they could arrive at the same places as men—if that's where they want to get to—*but doing it their way.*

References

Brown, P. T., & Dzendrowskyj, T. (2018). Sorting out an emotional muddle. *Developing Leaders* (29, Spring), 26–31. London: IEDP. https://iedp.cld.bz/Developing-Leaders-issue-29-Spring-20181/26/.

Bullmore, E. T. (2018). *The inflamed mind: A radical new approach to depression.* London: Short Books Ltd.

Damasio, A. (2000). *The feeling of what happens: Body, emotion and the making of consciousness.* London: Vintage/Random House.

Tokuhoma-Espinosa, T. (2018). *Neuromyths: Debunking false ideas about the brain.* New York: W. W. Norton.

Tokuhama-Espinosa, T. (2019). *Neuromyth: "Brain training" is supported by neuroscience.* https://psychcentral.com/blog/brain-training-is-supported-by-neuroscience/.

Wellness Creative Co. (2017). https://www.wellnesscreatives.com/fitness-industry-statistics-growth/.

4

Brain Sex-Based Attention and Communication

Kate says:

Attention ♀ Attention ♂

What we pay attention to determines what we notice about our world/ environment. This in turn determines what action we take in that environment. There are some key differences in what male and female brains pay attention to. In spite of our mosaic brains, whereby the blend of male and female attributes from both nature and nurture make us uniquely who we are, *there are* sources of difference that impact attention and communication in the structure and neurobiology of the brain—especially the way it connects information together. Understanding these differences and leveraging them is demonstrably good for business.

Combine leveraging generic brain sex difference with understanding and tapping into *individual* brain difference and you *really* do have access to all the brains in the business. Smart leaders do this. So first we look at what the general neurobiological female–male differences are during early child development. Then secondly, how they might show up in the world of modern business. You can then see how your own brain patterning fits with the typical neurobiological difference, your own particular brain sex and how this combination shows up for you in your workplace.

© The Author(s) 2020
K. Lanz and P. Brown, *All the Brains in the Business*, The Neuroscience of Business,
https://doi.org/10.1007/978-3-030-22153-9_4

Tuning in to Faces and Feelings

There is a developmental period called infantile puberty whereby the sex hormones in each sex influence the brains and bodies of the growing child. In boys this period lasts only nine months, in girls it lasts for twenty-four months with her body producing as much oestrogen (the reproductive hormone) as a fully-grown woman. Scientists believe (Styne and Grumbach 2002) that the purpose of infantile puberty is in order to set up the reproductive system for the future. Not surprisingly, the large amounts of oestrogen influence brain development in the infant female. The circuits for communication, empathy, gut feelings and the capacity to tune into the emotions of others are further enhanced at this developmental stage (Fig. 4.1).

Women's brains are primed from day one to notice micro facial expressions in others and determine from them what the emotions are that are being experienced by the other person. This is a skill that female babies develop from birth and it is true to say that, on average, women are better than men at reading emotion in the faces of others.

New born baby girls are born ready for mutual face gazing in a way that baby boys are not. In the womb, girls are not subject to the testosterone surge that turns the female embryo into becoming male and that shrinks the centres in the brain for communication and processing emotion. During the first three months of life, a baby girl's capability in eye contact and mutual face gazing increase by over 400% (Leeb and Rejskind 2004). In infant boys, there is no increase in face gazing capability at all during the same period. The skills of tuning into other's faces and seeking connection in this way

Fig. 4.1 Vase or faces? We all filter what we see differently (*Credit* Pio3/Shutterstock. com)

are hard-wired by nature. Social learning is not the determinant. Variations in social and cultural expression have not had time to establish themselves yet—gaze is a female–male difference.

Staying Connected and Seeking Approval

By nature women's self-defence responses differ from those of a man in a couple of key ways. First, the response to become aggressive under attack is less quick-fire in a female than a male (though a female defending her offspring will fight to the death). This is to do with her lower levels of testosterone and her stronger connections between the amygdala—the guardhouse of the brain—and the calming ruler of good judgement—the prefrontal cortex—than for a man (Campbell 2005; Giedd et al. 1996; Witelson et al. 1995). For our foremothers out on the savannah, a woman who was pregnant, nursing or caring for a young child could not have run or fought as easily as a man. Although none of us were there, so we are reasoning backwards, it is probable that women in our evolutionary history found safety in forming bonds, groupings and social ties that connected and protected them in threatening or difficult circumstances. There is therefore a fourth defence response in addition to 'fight, flight or freeze' that has been identified. This defence response is known as 'tend and befriend' (Taylor et al. 2000).

In a lecture on neuroscience, the speaker asked the group to imagine that they were with their tribe out on the savannah when a male from another tribe bursts in on the gathering in a distraught state. 'What would you do?' asks the lecturer. All the men in the room responded immediately saying, 'Kill him!' along, it has to be said, with a handful of the women. However, from in among was a chorus of female only voices saying, 'Offer him food and bring him by the fire and see what was wrong'. Clearly, this is not a statistically sound analysis but it was a neat illustration of women's 'tend and befriend' response. Supporting each other as a social grouping would have been a way to help ensure the survival of their offspring out on the plains of our human history. It is still a behaviour well observed in primate research.

This bio-evolutionary understanding needs to be connected to and tuned into the social grouping. A young girl's brain seeks social approval from others very early. She will be looking to make sense of the facial expressions and tone of voice of others to see how she is doing in relation to her connection to the group. In a study at the University of Texas (Rosen et al. 1992) of one-year-old boys and girls, the youngsters were brought by their mothers into a room with an object in it and every move was filmed. Each toddler was told not to touch the object once at the outset and then mum stood off to one side while the child was left to explore the room. Very few of the baby girls touched the forbidden object and they looked back at their mums between ten and twenty times more than the boys to seek for signs of either disapproval or approval. The little boys, on the other hand, hardly looked at their mums' faces and went off exploring the room, frequently touching the forbidden object even when their mums shouted not to. The little boys, with their higher testosterone levels and a larger (in men than women) part of the brain given to exploring their environment spatially (the temporal parietal lobe), had to have a hands-on experience of their environment without reference to the authority of their mothers.

Emotional Tones

Baby girls and boys hear differently. Female toddlers have been shown to respond to tiny changes in vocal tone. Girls can hear a wider range of sound frequency and tone in the human voice than boys can (Plante et al. 2006). The brains of new-born baby girls show a more heightened response to both their mothers' voices and to the cries of distress of a fellow infant than the brains of an infant baby boy (McClure 2000).

Research from the University of Sheffield (Sokhi et al. 2005) found differences in the way the male and female brains process voice sounds. Both men and women process voice sounds in Wernicke's area—the specialist area devoted to language in the left hemisphere. However men also process female voice sounds in the auditory section of the right hemisphere that is especially used for processing melody, while females tend to listen with both hemispheres and pick up more nuances of tonality in voice sounds and in other sounds (e.g. crying, moaning). So it seems that males tend to listen primarily with one hemisphere and do not hear the same nuances of tonality (e.g. may miss the warning tone in a female voice).

If wanting to know where Wernicke's area actually is, place the middle finger of your left hand against your skull just above and slightly behind the top part of your left ear. Then imagine going in through the skull and into your brain for about half your finger's length. You are now meddling with Wernicke's area (Fig. 4.2).

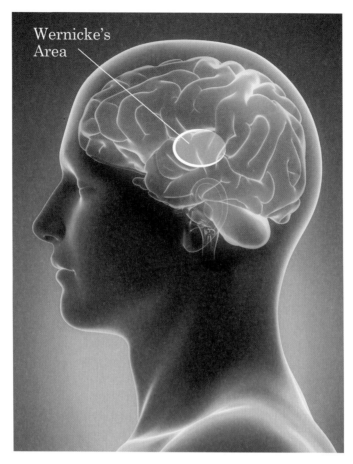

Fig. 4.2 Men process sound differently to women making them less able to distinguish tonal differences (*Credit* CLIPAREA I Custom media/Shutterstock.com)

Gut Feelings

During puberty the number of cells in a woman's body for tracking and making sense of body sensations increase. A combination of higher levels of oestrogen in the female body and larger areas in the brain tracking bodily sensations means that girls and women feel pain and gut sensations to a greater extent than boys do (Lawal et al. 2005; Derbyshire et al. 2002). Signals travel from the girl's gut to both her insula cortex—an ancient part of the brain that processes gut feelings—and her anterior cingulate gyrus—the part of the brain that judges, integrates and controls escape/avoidance emotions. On average, both these brain areas are larger and more active in women than in men (Levenson 2003)—as demonstrated in the meta-study referenced in Chapter 1 (Ruigroka et al. 2014).

Thus, the hormonal impact of oestrogen and the larger number of cells in the brain to process bodily sensations mean that, on average, a woman has a greater capacity to tune into gut feelings, and to sense and read other people's emotions than a man (Brody 1985).

Mirroring

There is a process called 'mirroring' whereby, as we watch someone else, especially when it is a loved one, our bodies respond like theirs, feels what they are feeling and our brain circuits light up matching what we are seeing. In a study at University College London (Singer et al. 2004, 2006), couples were tested in the fMRI scanner. The women went first and their brains were scanned as they received electric shocks to their hands ranging from weak to strong. The pain areas in their brains lit up to a greater or lesser extent. Then their beloveds were put into the same test situation. The women's brains continued to be scanned. They could neither see nor hear their partners. They were simply told about the strength of electric shock that their partner was receiving. The same brain areas lit up in the women when they were told their partners were being strongly shocked as had done for them when they experienced the pain first hand. Research has been unable to elicit the same response in men (Brizendene 2006).

Evolutionary biologists believe that this capacity to tune into another's feelings to such a remarkable extent had a twofold purpose (Campbell 2005). First, it would allow a woman to sense early any potentially danger-ous or aggressive behaviour from alpha males in her tribe and avoid it so as to protect her young. Second, it would allow her to understand the needs of her pre-verbal child to a minute extent.

Language—Spoken and Unspoken

Spoken language is processed in both hemispheres in the female brain and mainly in the left hemisphere for men. Emotion is mainly processed in the right hemisphere in both sexes. This means that for the male brain, a signal is not so readily connected to an emotion to be processed and then expressed verbally. In the brain centres for language and hearing, women have 11% more neurons than men (Witelson et al. 1995; Knaus et al. 2006).

On average, the female brain pays more attention more of the time, and is noticing and processing the emotional and relational context of a situation, given the way the connectivity works inter (versus intra) hemispherically, than the typical male brain. This pattern of brain activity, the larger areas for communication and the predisposition for connecting and collaborating, mean that little girls—and grown women at work—tend to speak more than little boys and men (Fig. 4.3).

Early speech patterns tend also to show gender difference. Deborah Tannen's studies of two- to five-year-old girls (Tannen 1990) noted that girls usually make more collaborative proposals using affiliative speech patterns than boys, starting their sentences with seeking to connect such as, 'Let's play…'. The little girls tended to use language to get consensus and influence without telling people exactly what to do. Research has shown that little boys know how to use this affiliative speech style but they don't tend to use it as often. Research has also demonstrated that boys tend to use language more often to command others: and that the boys frequently ignored comments or commands given by the girls. The impact of testosterone on the developing brain limits the male desire for spoken language and for connection but enhances it for competition and action.

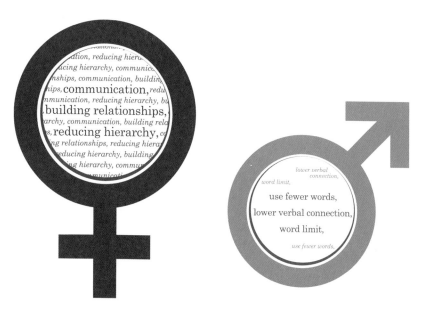

Fig. 4.3 Females tend to use much more speech than men (© Kate Lanz)

Empathy—The Magic of Intuition

If we take all of these differences together—the higher amounts of oestrogen and oxytocin (the bonding hormone) increasing her capacity to want to watch other's faces minutely; her superior hearing for micro changes in tone in the human voice; her more intense gut reactions and sensitivity to pain, including other people's and especially loved ones'; the greater circuitry to 'mirror' another's emotions, even without the visual connection; all combine to create a female brain that is highly tuned for empathy. Two million years of evolution seem to have primed the female brain to seek connection, collaboration, emotional understanding of others and a desire for social harmony. This capacity for empathy can land as an astonishing capacity for intuition at understanding the feelings of others in the blink of an eye.

This innate capacity is an incredible source of high-octane performance fuel in the essentially male-defined workplace that most women inhabit. Over-looked and under-used, with energy wasted in holding it in check, it has huge potential if brought into play—not least as a source of negotiating potency if leveraged intelligently.

Attention and Communication in the Workplace

These neurobiological differences of course show up for us as leaders in the world of modern business. In my coaching practice, I see and hear the nuances of sex difference showing up in leadership situations all the time. Most often these differences are not noticed or understood and thus the opportunity to leverage them is lost. The following case study, taken from an actual situation (with the individuals carefully anonymised), is a case in point of when and how the differences between the male and female brain at work can be a potent source of productivity.

Case Study—The Deal

Mark is the CEO of a large private equity business. His right-hand person, the commercial director, is a woman, Annie, smart, sharp and bold and younger than Mark by a decade. They have worked together for a number of years and thoroughly trust each other.

Many of their deals are long and complex negotiations. Mark is one of the toughest negotiators in the business. He is very alpha—a quality that comes in two flavours, alpha positive and alpha negative. There is a fine dividing line that represents the tipping point between the two positions. Alpha positive tends to represent the positive impact in the world of higher testosterone—a bias to action in the most direct efficient way possible. It involves a forthrightness that might not always consider the minutiae of the feelings involved in a situation and a more quick-fire mono focus on the task in hand. Mark veers largely to alpha positive. When he is stressed, however, the reverse is true. Alpha negative, at its worst, is bullying and involves coercing others into what the alpha wants to do without reference to the thoughts or feelings of anyone else involved.

Over the years, with our coaching and through sheer experience working together, they had developed a deep understanding of their different styles and different brains and had learned to leverage the difference. Annie often picked up micro signals in negotiations that Mark just simply did not spot. On one notable occasion in a lengthy negotiation, she sensed that something was not quite right with the Finance Director on the other side of a deal. She could not specifically say what it was that had alerted her to this but her intuition was telling her clearly that they needed to connect with him offline, outside of the negotiation and find out what was going on. Perhaps it had been something in his facial expression she thought, a slight tightening around his mouth when he sat back. She couldn't quite remember—but she just knew. Mark disagreed with her and his sense was that the FD was simply grumpy because he, Mark, was playing hardball on certain aspects of the negotiation. Annie was not convinced and would not settle. She told Mark that she was going to connect offline with the FD.

She was right. She called the guy up and shared her concerns with him. He was surprised that she had reached out, but grateful. He was dealing with some internal politics on his side that was putting him in a vulnerable position and he was feeling unsettled and anxious. Annie took him for coffee to get under the skin of his situation and how it felt to him. She could see the potential to change a part of the offer in a way that could potentially create a win-win at no deep detriment to the negotiations. The fact that she had spotted his minute discomfort and reached out to him impressed him greatly and cemented a level of trust in his relationship with her. Mark was surprised but pleased again at Annie's seemingly magical capacity to tune into feelings of which he was simply not aware. They started to debrief post meetings, sharing what they had each noticed and how it played into their handling of the negotiation. Mark's brain is more towards the male end of the brain sex spectrum—scoring 4 on the brain sex questionnaire (see Chapter 1). Annie has a more female brain—with a score of 14.

Summary

Articulating the essence of what it is to be a female leader in the world of work which is largely designed by men for men is a tricky thing to do. I find myself hearing it, seeing it and feeling it in my work with leaders both in the privacy of the coaching room and in the workplace supporting teams to perform. Bringing the underpinning neuroscience to the conversation such that the differences can at first be understood by men and women alike and enabling a new vocabulary to form around the new understanding is deeply empowering all round. This is not about which sex brain is 'better'. There is no such thing. We have evolved over millions of years to complement each other. The potency comes from seeking to acutely understand the differences between all the brains in the business such that any dominant culture does not inhibit the optimal activity of the brains within it, especially when they are counter-cultural. Even a man with a more female brain will have more testosterone than a woman and, due to his biological sex, will be far more likely to find a more male-biased work culture easier to fit in with than will a woman. Of course, there are always exceptions to the rule, everyone has their story of the female boss who was more alpha than most alpha males. That alas, is probably why she made it and a lot of what one sees in such circumstances, is, in my experience, trained in rather than innate for those types of women leaders.

One of my greatest joys in this work was when a female managing director, after the workshop on male–female brain difference, received an email from a male colleague containing a heartfelt apology. He apologised for having been somewhat dismissive in the past about the fact that he found her views on work issues slightly odd and certainly very different from his. He explained that he could now see why she came at things from such a different point of view and that he could with his new insights from neuroscience see how she brought a different vantage point to their work together. He committed to including her and listening to her in a very different way going forward so that they could better combine their different thinking. She was delighted. The relationship underwent a huge positive gear change. Men's and women's brains have a different overall 'take' on the world. Sustainable success into the future of work will depend significantly on leveraging this potential advantage that is so readily available.

References

Brizendene, L. B. (2006). *The female brain* (p. 164). New York: Broadway Books.

Brody, L. R. (1985). Gender differences in emotional development: A review of theories and research. *Journal of Personality, 53*(2), 102–149.

Campbell, A. (2005). Aggression. In D. M. Buss (Ed.), *Handbook of evolutionary psychology* (pp. 628–652). Hoboken, NJ: Wiley.

Derbyshire, S. W. G., Nichols, T. E., Firestone, L., Townsend, D. W., & Jones, A. K. P. (2002). Gender differences in patterns of cerebral activation during equal experience of painful laster stimulation. *The Journal of Pain, 3*(5), 401–411.

Giedd, J. N., Rumsey, J. M., Castellanos, F. X., Rajapakse, J. C., Kaysen, D., Vaituzis, A. C., et al. (1996). A quantitative MRI study of the corpus callosum in children and adolescents. *Developmental Brain Research, Elsevier, 91*(2), 274–280.

Knaus, T. A., Bollich, A. M., Corey, D. M. Lemen, L. C., & Foundas, A. L. (2006). Variability in perisylvian brain anatomy in healthy adults. *Brain and Language, Elsevier, 97*(2), 219–232.

Lawal et al. (2005). *American Journal Physiology and Gastrointestinal Liver Physiology, 289*(4), 722–730.

Leeb, R. T., & Rejskind, F. G. (2004). Here's looking at you, kid! A longitudinal study of perceived gender differences in mutual gaze behaviour in young infants. *Sex Roles, 50*(1–2), 1–5.

Levenson, R. W. (2003). Blood sweat and fears: The autonomic architecture of emotion. *Annal of the New York Academy of Sciences, 1000, 348–366.

McClure, E. B. (2000). A meta-analytic review of sex differences in facial expression processing and their development in infants, children, and adolescents. *Psychological Bulletin, 126*(3), 424–453.

Plante, E., Schmithorst, V. J., Holland, S. K., & Byars, A. W. (2006). Sex differences in the activation of language cortex during childhood. *Neuropsychologia, 44*(7), 1210–1221.

Rosen, W. D., Adamson, L. B., & Bakeman, R. (1992). An experimental investigation of infant social referencing: Mother's messages and gender differences. *Developmental Psychology, 28,* 1172–1178.

Ruigroka, A. N. V., Salimi-Khorshidi, G., Lai., M.-C., Baron-Cohen, S., Lombardo, M. V., Tait, R. J., & Suckling, J. (2014). Neuroscience and biobehavioural reviews. *Elsevier, 39,* 34–50.

Singer, T., Seymour, B., O'Doherty, J., Kaube, H., Dolan, R. J., & Frith, C. D. (2004). Empathy for pain involves the affective but not sensory components of pain. *Science, 303*(5661), 1157–1162.

Singer, T., Seymour, B., O'Doherty, J., Stephan, K. E., Dolan, R. J., & Frith, C. D. (2006). Empathic neural responses are modulated by the perceived fairness of others. *Nature, 439*(7075), 466–469.

Sokhi, D. S., Hunter, M. D., Wilkinson, I. D., & Woodruff, P. W. (2005). Male and female voices activate distinct regions in the male brain. *Neuroimage, 27*(3), 572–578.

Styne, D. M., & Grumbach, M. M. (2002). Puberty in boys and girls. In D. W. Pfaff, A. P. Arnold, S. E. Fahrbach, A. M. Etgen, & R. T. Rubin (Eds.), *Hormones, brain and behaviour* (pp. 661–716). San Diego, CA: Academic Press.

Tannen, D. (1990). Gender differences in topical coherence: Creating involvement in best friends' talk. *Discourse Processes: Special Gender and Conversational Interaction, 13*(1), 73–90.

Taylor, S., Klein, L. C., Lewis, B. P., Gruenewald, T. L., Gurung, R. A. R., & Updegraff, J. A. (2000). Biobehavioural responses to stress in females: Tend-and-befriend, not fight-or-flight. *Psychological Review, 107*(3), 411–429.

Witelson, S. F., Glezer, I. I., & Kigar, D. L. (1995). Women have greater density of neurons in the posterior temporal cortex. *The Journal of Neuroscience, 15*(5), 3418–3428.

5

Power, Politics and Pressure

Kate says:

'Be the best man you can be' is the way that many modern organisations recognise and encourage their leaders—men and women alike—to take their power. The trick that this approach profoundly misses is that the way women tend naturally to take their power is different from the way that men typically express and take their power. In my experience, the more female or male the actual brain in question, the greater the difference in the way power is expressed and used. An organisational culture that promotes ways of working that expresses power in a more typically male way is very likely to be under-leveraging many of the brains in the business—notably the more female ones. Our bespoke client research shows that this is indeed the case with up to 30% of the brains in certain key populations being underutilised because of work practices that value more typically male approaches to power and politics. This 30% includes men with higher female brain sex scores.

Hierarchy and the Male Brain

As detailed in Chapter 4, the male brain and body produce between five and twenty times more testosterone (T) than the female brain and body. In a man, the testosterone is produced from the testes; in the female it is the peripheral tissue to the ovaries plus the adrenal gland that produces testosterone. The potency of the testosterone is less in women for this reason. So,

© The Author(s) 2020
K. Lanz and P. Brown, *All the Brains in the Business*, The Neuroscience of Business,
https://doi.org/10.1007/978-3-030-22153-9_5

on the whole, women have less T and the T they do have has less of a kick than it does for the average male.

Testosterone, the macho king of the hormones, promotes the behaviour of dominance and competition, fuelling a focus on the importance of hierarchy and protecting one's turf (Tremblay et al. 1998). So even a low T man is likely to care more about his place in the pecking order than the average woman, or at least he may care more about being further from the bottom of the ranking than a woman for whom her ranking may not matter at all. This neurobiological fact has a huge impact on workplace culture. Let's take an even closer neuroscientific look.

Competition—The Neurobiology

Because of their significantly higher T levels, men tend to figure out their place in the order of things at work through competition. Research has shown that no matter how hard we try to influence (or not) our children, little girls tend to play house and care for dollies and little boys tend to race about fighting each other, or imaginary enemies, with whatever objects they are provided with (a plastic dinosaur makes a great gun I discovered with my sons). In general, boys are more interested in competitive games and girls are more interested in cooperative games (Maccoby 1998). Studies show that young boys spend 65% of their free play in competitive games and young girls only 35% (Knickmeyer et al. 2005). And the girls are twenty times more likely to engage in turn-taking behaviours than the boys.

In a nursery school study of boys and girls, the boys had unanimous agreement about the ranking of all the boys in the group by the end of the second play session. The rankings remained stable for the full six-month period of this particular piece of research. The girls, on the other hand, showed some social dominance but rankings were much more fluid. Research has shown that by the age of two, a boy's brain is driving him to establish physical and social dominance (Edelman and Omark 1973).

One study established that boy alpha leaders were not always the biggest but were the ones who refused to back down during conflict. The alpha boys all tested higher for T levels than the boys lower down the pecking order (Weisfield et al. 1987).

The scanning of boy brains playing computer fighting games, based on research from Stanford University, showed that the thrill of winning and beating opponents activated the reward centres in the brains of the boys. The greater the conquering of opponents, the greater the activation of the dopamine receptors in the male brain. Thus, competition provides a genuine neurochemical thrill (Weisfield et al. 1987).

Collaboration—The Neurobiology

Women are more likely to seek collaboration than most men. Why is this? On average, women produce between five and twenty times more oxytocin than men. Oxytocin is the bonding hormone and promotes collaborative behaviour—as illustrated in the research into girls in the playground showing that girls playing together are twenty times more likely to engage in collaborative turn-taking than their same age, male playmates. The girls' language patterns reflect this too. Young girls use collaborative language more often than young boys do. In her research, Deborah Tannen noted how girls of between two and five year's old used collaborative proposals by starting sentences with 'Let's play …' to get other girls to engage in a particular game. Her studies showed that girls used language more often to get consensus without telling others directly what to do. Research also shows that boys know how to use this affiliative speech style, but they use it much less than girls do. Instead, they more often use language to command others, boast or threaten (Tannen 1990).

With lower and less potent testosterone levels, women generally care less about their place in the pecking order and more about being in relationship with colleagues rather than competing with them. It is far more through collaboration and relationship that women get things done. So they would naturally activate and use their power in the organisational system, but they do it through relationship to a far greater extent than men (Baron-Cohen et al. 2003; Hoffman 1977; Davis 1994; Eisenberg and Lennon 1983) though from their T- driven perceptions of the world men are blind to those processes and do not value them as organisational behaviours. In a large bank, very predominantly male in both culture and numbers, a diversity and inclusion initiative that offered a programme to help create a coaching culture had no male takers at all and did not run, to the pained frustration of the three women who had spent a good deal of time and effort getting grudging acceptance that it would be worth doing.

This is one example among many of the fact that relational power taking is so often overlooked, not seen, and hence not valued in organisational cultures because they are geared up to reflect male-brained behaviours. This is not to lay blame. It simply is the way the male brain functions.

In my experience, this relational power taking is often overlooked, or used but not given reward, in organisational cultures being more geared to reflect more male-brained behaviours. Many large corporations say they want to create cultures of cooperation and collaboration to enhance innovation and agile execution. Yet they overtly measure, reward and encourage behaviours that promote individual winning, drive hierarchy and promote competition. This more male view of what power is and what success looks like exists in both the informal ways of behaving and in the formal performance management structures in most large corporates.

Case Study—Financial Services Recruitment

A client company in Financial Services recently undertook an analysis of their recruitment statistics. They are committed to diversity and inclusion and increasing the number of women they hire. However, recent hiring patterns were heading in the opposite direction and they were puzzled as to why.

I helped them to have a close look at their hiring procedures using the applied neuroscience of understanding gender. A key pattern that emerged was that during their selection process they were asking candidates to present a point of view on a case study. The (entirely male) panel then 'backed the candidate into a corner' to see how they reacted under pressure to defending their point of view. There was no 'right' answer to the case study. They simply wanted to see how well and for how long the candidate would hold and defend their position.

They ended up hiring far more men than women. The women were slower to come out guns blazing and most backed down far more often not by conceding defeat but by seeking a different option for on-going discussion. This was seen by the panel as a weakness under pressure by the panel. Everything about this part of their process was wrong from a brain gender point of view. They were making it literally impossible for themselves to see the best of the women in action (and quite probably some of the men with more female brains).

The women were being forced to work against their neurobiology in this situation. Relationship was being put at risk from the outset. In new circumstances, the female brain is far more likely to want to connect first, so being forced into an antagonistic position would set off all the survive responses in the brain. There was no chance of witnessing what these women were capable of. This testosterone-fuelled approach puts the higher oxytocin–lower T women at a disadvantage immediately.

Of course, in the real world, the successful candidate will experience situations where defending a position is important and where they will feel under pressure. The point is the way a female brain handles that type of pressure is different from the way a very male brain does. Oftentimes, in my experience, left to handle such a situation in a more natural way for that brain can create better longer-term business outcomes.

Matching Behaviours Mistaken for Low Confidence

Research in 2010 (Schumann and Ross 2010) demonstrated that women are significantly more likely to apologise during meetings than men. In fact, what women are doing in apologising is seeking to 'match' their colleagues and demonstrate equality or connection rather than competition and being higher up the hierarchy. 'Sorry if someone has already mentioned this, but one thought I've had is…'. The female intention is to put a colleague or opposite number in a meeting at ease. This tendency to equalise through the form of apologising is very often misconstrued as a lack of confidence. This is usually is not the case at all. I have had countless impressive, powerful women sent to my coaching consulting room where the seeming objective is to help the woman in question deal with a lack of confidence. The issue is rarely confidence. It is usually the case that there is in an incapacity to spot how these women are taking their power in a very different way. The compete vs collaborate brain preference is missed and causes confusion. The joy of this understanding is that women relax in relief as they realise that confidence is not the issue. They knew it was not. They felt unseen but could not put their finger on why. The problem was they were not being male. The neuroscience helps to shine a light on the conundrum. It also raises the possibility that many training courses for women leaders need to be radically re-shaped. The female brain manages its owner quite differently from the male brain. It is, after all, the brain's single critical job to manage its owner well not, when we are grown up, on the basis of a man's perception of how the world should be but in utilising its strengths most advantageously in context. Context is critical for getting the best out of all the brains in the business.

Being the best woman a woman can be organisationally is a great confidence booster, just as it is for a man to be the best man he can be. The Self ('I') then feels it is functioning at its best and the outcomes reinforce that sense of effective strength. For either a man or a woman, when doing their best on their own gendered terms gets serially invalidated, no wonder confidence goes down. But for a woman in a male world, the corporate blindness of the way she exercises her power is so often non-consciously endemic in the organisation that she is continuously invalidated. After losing the General Election that she unwisely called soon after becoming Prime Minister to lead the country into Brexit, Mrs. May justified her decision in answer to a reporter's question by saying: 'At least I had the balls to call it'. Whoever advised her to use such a phrase had absolutely no appreciation of the fact that neither men nor women would see such a remark as any kind of justification for her actions. Suddenly acquiring male characteristics is not a way to be an effective woman, though being an effective woman in a male world can be very tough indeed.

Office Politics—Workplace Culture and the Dominant in Group

Many of my clients complain about office politics and are worried that they are not very 'good' at it—men and women alike. I explain that it depends upon how you define 'office politics'. Humans are tribal. In our evolutionary history we are too weak to survive for long out on the plains alone. We have to exist in communities. As such, relationships are fundamental to our survival. Office politics is mostly about being in relationship with the various individuals with whom one is working. Office politics can be described along a spectrum from the benign day-to-day of interrelationship to the downright Machiavellian behaviours of those who would actively seek to stab you in the back. It is this Machiavellian end of the spectrum that understandably frightens people the most. This is the end of the spectrum that I call 'politics', the rest of it is simply skilful relationship management. Not surprisingly, due to brain differences women and men focus on different things in their relationship management. This is largely down to their brain differences.

An organisation that is serious about accessing the best of all the brains in the business must understand how power and politics play out differently for different sexed brains and make sure that the way power is recognised encompasses the best of both gender brains.

In most kingdoms, from the plains to the world of work, the dominant group defines the rules of the game and what gets recognised and rewarded. In any organisation that is the more likely to create strong 'in group' preferences for the ways of working that make up the culture. Research by David

Eagleman in the USA (Vaughn et al. 2018) put 130 participants in an fMRI scanner and showed them images of mixed-raced hands being stabbed by a hypodermic needle. The scanner looked at each participant's brain response to seeing a fellow human stabbed. The pain centres in participants' brains lit up as the needle entered each of the hands. This showed an innate empathy at watching another person's pain. Then the hands were given a label. Each hand was labelled a different religion and one hand was labelled atheist. The brains were scanned again as the hands got stabbed a second time. This time however, the predominant pattern was that there was no change in activation in the brain's empathy centres when members of the participant's 'out group' was stabbed. The majority of participants showed this non-empathic reaction to members of their 'out groups'. This even included the atheists. This reaction is innate, occurring way before conscious thought in the prefrontal cortex can kick in. These findings of what actually happens in the brain in relation to 'in group', 'out group' responses are of huge significance for creating brain-friendly work cultures.

The brain's limbic system is picking up what it feels like to be in the 'out group' pre-consciously and, when triggered, will automatically deflect attention (blood flow) from the cortex such that the quality of thinking in the cortex will, in part, become compromised. A prefrontal cortex that is not firing on all cylinders is not good for business. Cultures that inadvertently cause a lot of out-group responses are losing productivity and competitive advantage.

The Positive Impact of Office Politics

It is, however, important to engage in office 'politics' in terms of building relationships within the work eco-system. Relationships are what enable us to get things done in our work teams. Relationships help us learn as well as to feel safe and motivated at work. Research has shown that engaging in office politics has many upsides (at least at the end of 'politics' which is not the Machiavellian end). It reduces stress, enhances performance, advances individual reputations, facilitates career progression (King et al. 2018).

A 2013 study by the *Journal of Leadership Studies* (Westbrook et al. 2013) showed that both sexes see engaging in office politics as equally important. It also revealed that both sexes believe themselves to be good at politics, with women believing themselves as somewhat better at it than men. However, an Academy of Entrepreneurship study in 2015 (Phipps 2015) revealed for women and other minorities the benefits are not as great as for the men. This research asserted that it is not a matter of skill but rather of what gets recognised and rewarded. Why does this happen?

The neuroscience of 'in group' and 'out group' brain reactions, described above, is in part one of the explanations as to why the brain's non-conscious response accounts for this rewarding of in-group behaviours. 'Out group' behaviours are simply not noticed by the dominant 'in group'. The case study below is an exquisite example of how this happens.

Case Study—Going the Extra Mile Behind the Scenes

A new MD, recently in post, was determined to create a positive impact with her client and, in addition, show her company that she was worthy of the promotion. Late one afternoon, she called me in tears one for an emergency coaching session. What had happened to her was both a fascinating and perfect microcosm case in point about how the essence of what it means to be a woman in business and leadership in a dominant male culture can be completely overlooked by the essence of what it is to be male, in a dominant male culture. It was a microcosm of how office politics and power can play out at their worst at significant individual and organisational cost.

The company in question was helping a large client to develop and execute a new digital strategy. The new MD managed to get the client noticed sufficiently, internally at her company, to be invited to a large innovation event overseas and get their particular project up on to the stage as a case study to be shared at the innovation conference. This was real kudos for both the client and the MD's company. She was one member of a large team that had done the strategic analysis, design and were supporting execution. She was determined that the client would show up well at the prestigious event. She did not want to put herself front and centre of the public display on stage as she felt that this was not the place or time for herself personally to seek any glory. She worked diligently behind the scenes to make sure that all the people engaged in preparing and presenting the case study had all the relevant information they needed, that they were comfortable and confident with what they needed to do, that they were nudged along in a timely way to be ready, that all the internal stakeholders at her own company were in the loop and no one was taken by surprise. She had held the relational network clearly in mind for several weeks and leant in making calls, sending mails, popping in to see people—all additional tasks on top of her day job. She was determined that the show would go well.

The day came and the client, along with a couple of people from the MD's own company, did a stellar job presenting the project. At the end of the presentation the new MD's male boss took to the stage and made a great show of saying thank you to the team who had made the work and the presentation such a significant success. He did not include her at all in his thanks. She felt utterly devastated and deflated. She realised that he had not noticed ANY of what she had been doing for weeks. He was not an unpleasant man at all, in fact, she really liked him and thought he was a good boss. She was stunned at the oversight. She had a sleepless night that night and felt demotivated for a good couple of days. Unable to reignite her mojo, she called me up on day three to debrief.

Such was the impact that it had even occurred to her to resign. She had recently been approached by a competitor company. The cost to her employer to get her through to MD level ran into the hundreds of thousands of pounds over the years she had been there and she was on the top talent grid. She was not someone the company wanted to lose. Now she was considering leaving, such was the depth of her 'survive' response to the casual oversight—and she had not worked at 100% capacity for three days.

In our coaching session we explored the male–female brain and behaviour differences that this event had highlighted. We ended up with her considering how she could leverage the incident to engage the leadership community above her in understanding the brain-sex differences that had been at cause to create a dialogue about what might be changed in the ways of working that would impact the work culture in a positive way. She took power from her difficulty to create positive change.

The Neurochemical Double Bind

Research by Kate Davey in 2008 shows that women associate political behaviour with more traditionally masculine ways of going on and with which they do not feel comfortable. The differences between male and female brains explain why. Male higher testosterone levels mean men are generally happier by nature to compete politically, especially within cultures that have been designed largely by men for men. Davey's research shows that, for women, playing the political game comes at a cost. This is not surprising as women with higher oxytocin levels naturally have a stronger tendency, on average, to collaborate rather than compete. Mrs. May's approach to her male European counterparts in the Brexit negotiations was always to believe that dialogue was possible, while they wanted to position themselves as being the controllers of the situation.

In Davey's research, women reported lower individual motivation and increased stress as a result of having to play the politics in a more male way. In the stress response for women, oxytocin levels go up, increasing the tendency to resolve the stress in a non-competitive way (tend and befriend). For men under stress, testosterone levels go up and oxytocin levels go down. This neurochemical female–male difference goes a long way to explaining Davey's (and many others) findings. It would also be interesting to understand how many men with more female brains found the same negative side effects; and how women with more male brains fared.

Summary—The Testosterone Conundrum

Kate Davey's research into office politics highlights how those processes are part of the informal system that keeps power with those who have it within a culture. Breaking into the old boys' club is tough. Ask just about any female executive who's tried. Lots of companies make valiant attempts to raise awareness about the 'boys' club' dynamic and bring it into the open and conscious awareness. These efforts are worthy. They do not, however, change things very quickly or sustainably. That is a neurochemical conundrum involved with testosterone at its heart. This silverback alpha of the neurotransmitters does not want to give up power and is a large part of the driving force that makes up the informal power systems that keep the power with those who have it. As we have seen, men by nature are more concerned with their place in the pecking order. Women take their power differently, more through relationships and collaboration. This testosterone–oxytocin difference is much the cause with the difficulty in changing the informal power structures and putting women in the position of having to behave like men to get on within them. The very subtle undercurrents in the work culture that this creates is a significant cause in the loss of access to the power of all the brains in the building. We are hopeful for the formulation we have arrived at in the Introduction—that e-quality is a future that both men and women can agree about, with it being a joint commitment, not a competitive task. The emergence of feminine energy into the organisation is the prize to be had.

What will it take to get there? It will take men and women working together in true partnerships at all levels across an organisation to identify the moments when the testosterone conundrum is getting in the way of accessing the full and true power of all the brains in the business. One of the most potent forces for good in this regard that I have come across are senior men in their 50s who have daughters entering the workforce after finishing university. As these powerful men start to confront the reality of what their daughters will face, they start to take positive action towards positive discrimination. Another key factor in their desire to create positive change for women is that their own testosterone levels are significantly on the decline so their desire to compete is significantly reduced when compared to their twenty- or thirty-something selves!

But we think that positive discrimination is not the effective answer (there is another form of discrimination which works far better and we come to in Chapter 6). It implies the righting of a wrong and triggers men's competitiveness, neither of which are desirable in context. It also has the effect of seriously disturbing real meritocratic structures. It is finding a shared delight in what the differences are and what added value they bring that is the approach that will make for sustainable change: and we think that e-quality is what it is about.

Men are in the huge majority in leadership positions in large organisations. Senior men with the power to change their organisational systems for the better can be even more proactive in supporting their female colleagues to be enabled to take their power as women. Female leaders need to leverage and speak up about the differences that the female brain brings to all aspects of corporate decision-making, in the full knowledge that it is that very difference that has particular value to add.

References

Baron-Cohen, S., Richler, J., Bisarya, D., et al. (2003). The systematising quotient: An investigation of adults with Asperger Syndrome or high functioning autism and normal sex differences. *Philosophical Transactions of the Royal Society, Series B.* Special issue on Autism: Mind and Brain, *358*, 361–374.

Davey, K. M. (2008). Women's accounts of organisational politics as a gendering process. *Gender Work and Organization, 15*(6), 650–671.

Davis, M. H. (1994). *Empathy: A social psychological approach.* Social psychology series. Colorado: Westview Press.

Edelman, M. S., & Omark, D. R. (1973). Dominance hierarchies in young children. *Social Science Information, 7*(1), 103–110.

Eisenberg, N., & Lennon, R. (1983). Sex differences in empathy and related capacities. *Psychological Bulletin, 94*, 100–131.

Hoffman, M. I. (1977). Sex differences in empathy and related behaviours. *Psychological Bulletin, 84,* 712–722.

King, Denyer, Party. (2018). *Organisational Dynamics, 33*(4), 2004, Perrewe and Nelson, *Harvard Business Review,* September 2018.

Knickmeyer, R. C., Wheelright et al. (2005). Gender typed play and amniotic testosterone. *Developmental Psychology, 41*(3), 517–528.

Maccoby, E. E. (1998). *The two sexes growing up apart and coming together.* Cambridge, MA: Harvard University Press.

Phipps, S. T. A. (2015). Women versus men in entrepreneurship: A comparison of the sexes on creativity, political skill, and entrepreneurial intentions. *Academy of Entrepreneurship Journal, 21*(1), 32–43 (Middle Georgia State College Leon C. Prieto, Clayton State University).

Schumann, K., & Ross, M. (2010). Why women apologize more than men: Gender differences in thresholds for perceiving offensive behavior. *Psychological Science, 21*(11), 1649–1655.

Tannen, D. (1990). Gender differences in topical coherence: Creating involvement in best friends' talk. *Discourse Processes: Special Gender and Conversational Interaction, 13*(1), 73–90.

Tremblay, R. E., Schaal, B., & Boulerice, B. (1998). Testosterone, physical aggression, dominance, and physical development in early adolescence. *International Journal of Behavioural Development, 22*(4), 753–777. First Published December 1.

Vaughn, D. A., Savjani, R. R., Cohen, M. S., & Eagleman, D. M. (2018). Empathic neural responses predict group allegiance. *Frontiers in Human Neuroscience, 12,* 302.

Weisfield, G. E., et al. (1987). Stability of boys social success among peers over an eleven-year period. *Contributions to Human Development, 18,* 58–60.

Westbrook, T. S., Veale, J. R., & Karnes, R. E. (2013). Multirater and gender differences in the measurement of political skill in organization. *Journal of Leadership Studies, 7*(1). https://doi.org/10.1002/jls.21275. Accessed online 11 April 2019.

6

Problem Solvers and Solution Seekers—The Difference Between Intra-compared with Inter-hemispheric Connectivity

Kate says:

In a ground-breaking study in 2013 (*PNAS* 2014) researchers from the University of Pennsylvania (U Penn) demonstrated that the typical male brain functions within each hemisphere—with the neural connections firing front to back of the right or the left hemisphere independently. There is evidence that this neural connectivity leads to a more systemising, singular focus with a preference for binary solutions—'either/or'. This is highly effective for solving certain types of problem. The U Penn research demonstrated that women's brains, on the other hand, typically function by integrating input from both hemispheres concurrently—'both/and'—facilitating women's understanding and capacity for continuous, iterative steps towards emerging solutions. The researchers describe this as 'female brains are designed to facilitate communication between analytical and intuitive processing modes' (*PNAS* 2014, p. 823) (Fig. 6.1).

As we saw in Chapter 1, this shows the intra-connectivity (male/blue) and inter-connectivity (female/orange) patterns in male and female brains.

These two very different operating modes are generally not understood in business and, in consequence, their complementary advantages are massively underutilised in a majority of organisations. My bespoke company research is showing that there are large pockets of underleveraged brain power sitting in the offices of today's businesses. In the worst cases, the lack of understanding of the ***difference*** between the brains means that a high percentage of the potential of good brains is being actively, though unintentionally, switched off during the course of the working day. This makes no sense

© The Author(s) 2020
K. Lanz and P. Brown, *All the Brains in the Business*, The Neuroscience of Business,
https://doi.org/10.1007/978-3-030-22153-9_6

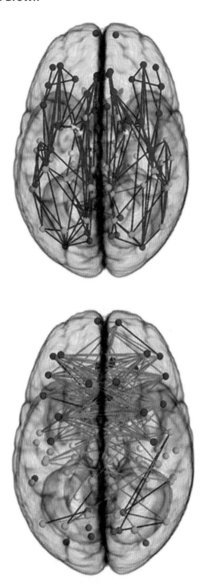

Fig. 6.1 The intra-connectivity patterns in the male (blue) and inter-connectivity patterns in the female (orange) brains (*Source* Ingalhalikar et al., *PNAS* 2014 January, 111 (2), 823–828. https://doi.org/10.1073/pnas.1316909110)

from the point of view of maximising the return on investment on *all* the brains in the business. It is the workplace equivalent of driving a car with the handbrake on.

Much of the argument we see in relation to the modern gender agenda is a demand for equality. It is time that this demand for equality needs to be replaced by a proper understanding of ***the value of the difference***. Based on the science behind the difference, this chapter shows how actually tapping into the differences can ignite thrive in the different gender brains, thereby escaping from the more common-place state of survive that has become accepted as the norm that so many working lives and so many 'engagement' surveys show.

The Problem-Solving Male Brain

The intra-hemispheric neural connectivity patterns in the male brain shown by the researchers at U Penn suggests a brain specialisation that facilitates 'connectivity between perception and coordinated action' (*PNAS* 2014, p. 823). One conclusion that the research leads to is that this intra-hemispheric neural connectivity enables the male brain to focus deeply on understanding and 'fixing' problems in a systematic manner. Simon (Baron-Cohen), Professor of Developmental Psychopathology at Cambridge University, has called this type of problem-solving 'systemising'. Baron-Cohen defines 'systemising' as 'the drive to analyse, explore and construct a system. The systemizer intuitively figures out how things work, or extracts the underlying rules that govern the behaviour of that system' (Baron-Cohen 2003).

This predisposition towards tasks that indicate the more male tendency to 'systemise' is evidenced through studies that show that, when given a choice of toys between dolls and building blocks/mechanical toys, boys as young as two years old will gravitate towards the blocks and mechanical objects (Baron-Cohen 2003). This preference for the mechanical over the human has been evidenced in even younger children at age one. Male one-year-olds will watch a film about cars for longer than girls of an equivalent age. Even day-old, infant male babies stare for longer at a mechanical mobile than a human face (Connellan et al. 2001; Lutchmaya and Baron-Cohen 2002). Studies of drawings from pre-industrial societies have also shown that boys more often depict mechanical type objects such as tools or weapons. All of this evidence indicates that the more male tendency to 'systematise' is innate rather than socially conditioned in modern society.

Naturally, being able to solve problems quickly and take a heavily task-based approach to the issue in hand has many advantages in the workplace. There are many areas of work and types of role where this is exactly what is required much of the time. It is perhaps partly as a result of this combination of 'systematising' focus and greater need for status in the hierarchy (fuelled by higher T levels than women) that men are on the whole more dominant in number and position in most organisations at present. Executive-capable women have only recently come into the workplace in any numbers and been able to rise to senior positions. This has been one of the many important wins brought about by the feminist movement over the last sixty years. It is also as a result of the efforts of the feminist movement that proper provision for maternity conditions supporting a career have begun to exist. In some countries, such as Sweden, the conditions for enabling successful parenting for women and men alike is impressive. In other countries, such as the US, there is still a hugely long way to go in this regard. But let's keep in mind the central proposition of this book—*Women can arrive at the same outcomes that men can create, or they might arrive somewhere else, but they would do it differently. Harnessing the difference and knowing how to do this is of central importance in the way forward.*

The Collaborative Female Brain

Women's neurobiological responses are, by comparison, generally more focused towards the human and relational side of an activity, interaction or task. A look at the underpinning neurological response in the brain shows that women more naturally focus on the human element. Another fascinating study exposed a group of adult men and women to a flow of different images comprising mechanical objects and human figures. The images were projected in quick succession into the exact same part of the visual field using a stereoscope. The recall of the two sexes was then tested. It revealed that men recalled seeing more mechanical objects and the women remembered seeing more human figures (McGuinness and Symonds 1977).

This indicates where female attention is more likely to go. What does this mean though? Women on average have a greater capacity to empathise spontaneously than most men. Baron-Cohen defines empathising as 'the drive to identify another person's emotions and thoughts, and to respond to them with an appropriate emotion' (Baron-Cohen 2003).

Women clearly need to be able to read emotion well to be able to empathise. Studies show that women possess a superior ability to read emotion in others, even through minute facial expressions, than most men (Buck et al. 1972; Baron-Cohen et al. 2001). They also score more highly, on average, on their capacity to respond emotionally appropriately (Hoffman 1977; Eisenberg and Lennon 1983).

Many similar research studies that show that, on average, males display greater interest in how mechanistic systems work and, on average, females display greater interest in and attention to the human, intuitive side of the equation.

Going Out for Dinner or Staying at Home!

These differences can be accounted for in part by our evolutionary history. It was the men who more often went to hunt and kill to feed the tribe meat while the women stayed back near the commune with the babies and engaged in the gathering tasks as a group. They spent more time caring for the young and generating plant-based food provision by this gathering. So men had to figure out how to track prey, often in silence, understand the territory in order to do this, craft the weapons that would allow them to kill. The act of 'killing the woolly mammoths' was the domain of the men. This behaviour requires more systemising skill than empathising skill. Whereas the women back at the commune needed to collaborate and connect with each other and keep a close eye on and nurture their babies.

The very different skills required for these two very different roles and activities are assumed by evolutionary biologists to play a role in the very different neural connectivity patterns discussed in Chapter 1. Intra-hemispherical differences, as well as inter-hemispheric differences, in neural connectivity increase during adolescence and into adulthood, indicating that our brain differences are hard-wired by nature.

A Woolly Mammoth in the Workplace?

While it is not the focus of this chapter to review these studies at length, the accumulated evidence all reinforces the same conclusions. It would not be accurate though to say that men are exclusively better at mechanical systematic tasks and women similarly at reading the emotions in the room and making the relational connections. However, as we have seen, and this research continues to highlight, ***there are, on average, differences between the sexes based on differences in brains along a male–female continuum***. And it is profoundly clear, from both science and experience, that these differences show up in how male and female behaviours are (mis)understood, displayed and (under) valued in the workplace.

Valuing the Difference

As we saw in Chapter 1, the sex of the brain is not necessarily the same as the sex of the body or appearance by which we socially type each other as male or female. Most of us have a mosaic brain with a combination of both male and female traits. Within each of our own brains we have countless individual differences that make us each the unique person that we are. The special quality of highly effective leaders is to be able to tune in and ignite thrive in the individual brains in their businesses. This is what underpins the art of highly effective leadership and, in turn, creates sustainable organisations.

The real power in all of this comes from accessing the significant and important differences between men's and women's brain function and accessing the complementarity that nature has spent millions of years evolving. And that raises the question of how organisations can help men to accept what women already know—that women's brains bring a different kind of understanding of and approach to the world; and that it might be of great benefit to organisations to leverage that fact.

Powerful Discrimination—A New Concept for Accessing the Best of the Brains in the Business

What we want to share with you is how these brain differences can be intelligently accessed at work. It is *only* by *powerful discrimination* between the different types of brain that *all* the latent brain power in the business will be accessed.

Powerful discrimination involves several things. Firstly, start from the awareness that mens' and womens' brains *are* different. Secondly, tune in to the different individual brains that are involved in any given task at any given time. Since along with the innate male–female brain differences, we have seen how brains exist along a brain-sex spectrum. Thirdly, deliberately seek to access the difference *on purpose* and do not allow the brain power present to go underutilised. Do not settle for less. This can be done in part by deploying some work practices that are designed to take account of and enable all the brains to feel validated, seen and heard. We look in much more detail into this in Chapter 8. Finally, don't leave a task without being sure that all the brains present are in thrive. If they are not, find out why.

Do not settle. And importantly remember that being in thrive is not about all agreeing with each other and neither does it mean false harmony. It is about skilfully insisting that what each sex brain brings to any piece of thinking is truly heard and that the person feels validated and enabled to contribute to the issues in hand.

This approach to powerful discrimination between brain gender difference is not yet commonplace in many organisations and we would like to help change that so that brains, the people they belong to and businesses all benefit.

What Underpins the Value in Powerful Discrimination?

Connectivity in the female brain works across both hemispheres concurrently as we have seen in Chapter 1. This pattern of neural connectivity means that the female brain is tuned into a broad focus in relation to the task, encompassing the relational and emotional to a greater extent on average than most men. This broad focus combines with the impact of higher levels of oxytocin driving a greater intrinsic desire to collaborate and connect. The larger communication centres and greater connection between analytical and intuitive processing modes (including gut feelings) means that women are more likely to take a broader, multi-dimensional approach to everything. This can often *appear* slower in moving the action forward. The brain that is using both hemispheres and pulling in information about the emotional and relational context is, to a greater extent, a brain that likes more naturally to iterate around the problem in hand, taking into account many facets of the problem, some of which are emotionally just hinted at for the woman, but with each iteration something different bubbles to the surface.

This iterative, intuitive process can frustrate a more male-brained man. Defining the world in a male way, he finds it inefficient and verbose, especially when his brain can see his highly direct route to fixing the problem. Along with the frustration this causes him is also the likelihood that he might care less than her about the feelings of all those involved—not in a callous way, but simply with higher testosterone levels, lower oxytocin levels and a brain that likes to lock on and solve the immediate issue, with a greater drive to 'fix it now'. After all, this is business isn't it? We've found the mammoth, cornered it, let's just kill it!

Truly allowing both brains' approaches to have the airtime and means they both need to facilitate them working together is the art and the neuroscience of great leadership. Knowing when to deploy powerful discrimination will give organisations the competitive advantage.

Here is how one client solved what, for him, was the problem of enabling female/male brain styles to work together to create win-win-win (which is itself a very male concept, but we will keep it for the moment as too much change at once is generally not healthy!).

Woolly Mammoths Are Easy to Spot—A Case Study

A senior leader heading up the diversity and inclusion effort at the global multi-media company where he had worked for many years said, 'The trouble is, Kate, we reward the folk who go out and kill the woolly mammoths round here. What we don't often notice is what's happening to enable them to do that. Round here most of the mammoth killers are men and the glue that holds it all together to enable them to go out and hunt comes from many of the women in the organisation. The women are not recognised for what they actually do to make this happen. Nor are they fully rewarded for it, which frustrates them, and too many good women go'. He was on a mission to change this.

It was good to hear a senior male member of the organisation acknowledging this but, as a company, they were finding it hard to create change on a number of fronts. They knew that without the so-called 'glue', the mammoth killers could not bring home the business, yet the support was not measured or rewarded. They also knew that the performance review system was more likely to motivate people who were visibly competitive and wanted to get out and get hunting. This type of activity tended to attract higher testosterone individuals and, unsurprisingly, most of them were men.

Nevertheless, this was an organisation that was aiming to make changes and wanted to understand how to do this. One such change was to appoint a woman, considered somewhat counter-cultural, to lead a large division where technology delivery was a fundamental part of the service provided by the business to its clients. Sandra was duly appointed—she had a big idea for the division and decided to build a new offering that would provide clients with an interdisciplinary team that could offer fast technical solutions for their business challenges.

Sandra appointed a highly experienced man, Doug, to lead the new initiative. He had shown real promise in a previous role and had been in the business for about six years. But in the early weeks into the role it became apparent that he was not showing the strategic flair she had hoped for, although he was delivering solid infrastructure and meticulous processes to support the foundations of the new function. She was disappointed and frustrated by the fact that he seemed unable to deeply engage with the big picture and create the buzz about the future scenarios that could emerge from what the division was trying to develop. He was also so focused on creating the platforms and processes that he was not connecting widely with senior stakeholders inside and outside the business. His presence was not being sufficiently strongly or widely felt.

Doug, on the other hand, felt as if he was drinking from a fire hydrant. There was so much detail necessary to set up the fundamentals upon which a whole new approach was to be built. He was very clear about the building blocks in structure, objectives and the types of people needed that would be the foundation of the new initiative. He worked flat out to move things forward fast. Part of the reason he was so 'head down' was that he just did not have the people on the ground to delegate to yet. Job number one was to build a team—fast.

Frustrated with the situation, Sandra consulted with one of her senior male mentors. His advice was that, were it down to him, he would remove Doug now from the role. He felt there had been too little visible strategic impact quickly enough and felt that Sandra should not waste any further time, find someone else to take Doug's place, and simply put Doug back where he had come from in the organisation.

Sandra brought the dilemma into coaching. She could not pinpoint exactly *why* she felt it, but her strong intuition was to stick with Doug and try a different tack with him. Sandra was acutely aware of the shame (a primal survival emotion, see page *About This Book*) being moved out of the new role after such short tenure would cause Doug and also the fact that it could quite possibly lead to a career dead-end within the organisation. She was not prepared to do this to him—or compromise the Company, if he chose to take his skills, and her vision for the department, elsewhere. So she took a more relationship-based view of the situation, determined to preserve Doug's confidence but also increase his impact, and decided that she was prepared to take some of the political 'heat' upon herself by running counter to some big opinions in the company.

Under the stress he was feeling, Doug had become even more highly task-focused. His attention to relationships and the broader, bigger picture had dropped further. He had locked on to the task of putting in place the fundamental building blocks to the exclusion of all else. He was inadvertently distancing himself from key relationships, including Sandra, and unaware of the impact that his responses were having on others around him.

Sandra could see, with some insights provided through coaching, that Doug, who had a high male brain sex score, and so a brain well on the male side of the spectrum, was not going to be able to do the bigger relational and strategic piece until later on in the process. Under the kind of pressure he was experiencing, his particular male brain could not settle until he had put what he saw as the foundations in place. Sandra knew this as she was aware about the neuroscience underpinning the male/female brain differences especially during stress. So Sandra decided to work with him in the way that his brain needed to be operating at that moment in time, determined to access the best of brain difference. She gave him clear, direct feedback arising from her perspectives about what was working and what was not working, taking the view that there was no point in sugar-coating (see more on feedback to different sex brains in Chapter 9). What she did was a very good example of the notion of powerful discrimination in practice.

Sandra enabled the context where they co-created a clear plan of the stages he would need to go through to deliver the service they wanted to offer to clients. This approach reduced his stress about the areas he realised that he was not attending to effectively at present. Sandra took on some of the key aspects of these areas in the meantime. One such aspect was engaging with particular key stakeholders and building central relationships that would be important as the product came to market. Sandra partnered with Doug, leveraging her different strengths to complement his. She was also careful to share with him what she was doing with building key relationships, how she was going about it, so as to build his awareness and skill in this area.

This enabled Doug to relax and get on with being effective at putting in the infrastructure, skills and processes that were pivotal. His stress levels went down and his effectiveness and pace went up. She was careful to publicise this to some of the big voices internally who had developed a negative view of him.

Over the months ahead, they continued to work in tandem in this kind of way, leveraging the best of both their brains, each at the more male and female ends of the spectrum respectively. This was soothing for both of their limbic systems. As their respective stress levels went down it helped them to build trust with each other. Doug's confidence continued to build and he started to become more effective in the areas that Sandra had felt were initially missing.

Sandra's leadership reputation was enhanced. A lot of people had witnessed that Doug's appointment had got off to a bumpy start but the fact that Sandra had stood by him in the face of political pressure bred confidence in the Division and increased the feeling of psychological and emotional safety across the piece.

This situation is a particularly clear example of how understanding male–female brain differences, especially under pressure, can save a situation and build productivity at many levels. Sandra's solution was not the most direct and seemingly efficient in the early stages of the process. She iterated on the issue several times, taking counsel from her senior mentor, her coach and soundings from people close to the situation. She took the time to tune in to her own intuition. She chose an approach based in the relational and emotional while still driving towards the business outcomes she wanted. She used brain gender knowledge to good effect and in doing so saved Doug's career, as well as building levels of trust across her division and getting the job done. A company-wide engagement survey that just so happened to be running at the time showed some of the most positive results across the company with people feeling motivated and valued in Sandra's area. People felt safe and trust was high, enhancing Sandra's ability to drive for high-performance. A winning combination.

Accessing the Best of Both Sex Brains— The Gendersmart Solution Spectrum™

From everything we have reviewed over the previous chapters about brain gender, we can see that there are significant and important female–male brain differences that cause different behaviours and to which many organisations are blind. These differences can be translated into typical behavioural distinctions which discriminate between female and male brain function.

We have developed a model, the Gendersmart Spectrum™, that distils these behaviours into five spectrum distinctions. While recognising that any model is, by definition, a significant simplification of the subtlety which is the reality of an individual brain, this model can help busy leaders fast track their way to more skilful partnering between different sexed brains. The following five distinctions are key to understanding and accessing the best of male–female brain sex difference:

Gendersmart Spectrum™ (© Kate Lanz)

Single fastest route to fix the problem	Iterate for emerging solutions
Mono task focus	Hold multi possibilities
Transactional interactions	Relationship focused interactions
Low emotional context	High emotional context
Silo—separate stance	Collaborative inclusive stance

To make sense of the model, we need to look at each distinct behaviour on its own spectrum. We each have preferred ways of achieving things in the world and approaching issues and challenges; which very much includes the way that we relate to and interact with people. We respond to others or demands made on us somewhere along each of these spectrums. All of our personal preferences in this regard are unique to us. They exist as a bespoke combination of neural pathways that will fire up and cause us to respond in a particular way given our combination of brain sex—by nature—and experience growing up in the world—giving us our brain sex by nurture.

What people say to us, how they say it and our immediate perception of whatever are the implications of what we hear gets a brain response in 85 milliseconds (see the preface on our eight emotions). That's our limbic system, testing everything all of the time for its emotional significance. That's fast. Making conscious sense of this information comes at 250 milliseconds, through the filter of the prefrontal cortex (the CEO of our brain) to put it all together and bring language into play. That's relatively slow. The more tuned in we are to our own triggers for survive and thrive, and to those of the people we work with, the better we can communicate and partner with our colleagues. As we saw with Sandra and Doug, it can take a lot of conscious effort to arrive at a mode of working that allows different brain types to stay in thrive. Sandra worked hard to keep in touch with where she and Doug were on the spectrum of each behaviour in order to generate the right conditions for both of them to work to their best. A leader needs to be able to recognise feelings, which come from the radar scanner of the amygdala, to attach thought to those feelings in order to create intelligent emotions—emotions in the service of thought but, paradoxically, only properly understood through thought. And then for these thoughts and understandings to be translated into a modus operandi for working.

The changes you might need to make could be quite subtle but therein lies the power of this model. Our limbic systems responds within 85 milliseconds to any stimulus that triggers one of the eight primal emotions. Even a small change from you that avoids triggering a survive emotion in the other person, will save you significant amounts of time and energy and enhance productivity.

Sandra and Doug operated predominantly on different sides of the Gendersmart Spectrum™ and their differences became accentuated under pressure. What they were able to do as a result of Sandra's courage to trust all of the data she had in her feeling system was to deploy the positive differences to deal with the complex task in hand, and use the fact that they were working to their brains' strengths in the early phase of the project to help reassure them and reduce doubt (and so survive feelings). Then they could build trust and ultimately partner together to get the job done. They did this in a way that had a big positive ripple effect across the Division. Trust rippled outwards as a significant by-product of the way they managed to interact. The situation could have turned out very differently but Gendersmart™ brain partnering helped them turn it around.

Summary

Iain McGilchrist, fellow at Oxford University, psychiatrist, neuroscientist, argues in his book, *The Master and His Emissary* (2009), on the basis of much neuroscientific evidence, that the remarkable material success and achievements of the western world over the past three centuries has been due to the increasing social dominance of the left brain; but that it has been done at the price of diminishing the social importance of the creative, speculative, imaginative right brain. The left brain, he argues, narrows attention, de-personalises and restricts everything to observable outcomes but misses out on the bigger picture. The right brain looks for meaning and purpose and possibility in life, and welcomes ambiguity and uncertainty.

We, as a species, face some of the most major and far-reaching issues, such as climate change, water and energy supply shortages and antibiotic resistance, to name some of the key challenges. Finding solutions will require immense collaboration across many disciplines, and as Iain McGilchrist has identified, perhaps by allowing the right side of the brain further freedoms to explore possible solutions, in tandem with what the left side has to offer makes a great deal of sense.

The argument we want to make is for *powerful discrimination* as a positive force for accessing the difference between the female and male brain. Equal pay for equal work is a must and modern efforts to redress the unacceptable imbalance need to continue. However, the notion that women and men are equal is missing a potent source of productivity and joy. Over millions of years, women and men have evolved in ways to **complement** each other for the survival of the species.

Brain science is supporting us to understand the differences more clearly. It is well researched and understood that when we trust, we thrive. The brain then functions at its creative best. Work environments that enhance the capacity for brain differences to be a welcome phenomenon and actively enable partnering between different sex brains will win in business. So we look next at the business case for brain gender breakthrough.

References

Baron-Cohen, S. (2003). *The essential difference: Men, women and the extreme male brain* (p. 3). New York: Basic Books.

Baron-Cohen, S., Wheelright, S., & Hill, J. (2001). *Journal of Child Psychology and Psychiatry, 42,* 241–252.

Buck, R. W., Savin, V. J., Miller, R. E., & Caul, W. F. (1972). Communication of affect through facial expression in humans. *Journal of Personality and Social Psychology, 23*(3), 362–371.

Connellan, J., Baron-Cohen, S., Wheelright, S., et al. (2001). Sex differences in human neonatal social perception. *Infant Behaviour and Development, 23,* 113–118.

Eisenberg, N., & Lennon, R. (1983). Sex differences in empathy related capacities. *Psychological Bulletin, 94,* 100–131.

Hoffman, M. L. (1977). Sex differences in empathy and related behaviours. *Psychological Bulletin, 84,* 712–722.

Lutchmaya, S., & Baron-Cohen, S. (2002). Human sex differences in social and non-social looking preferences at 12 months of age. *Infant Behaviour and Development, 24*(4), 319–324.

McGilchrist, I. (2009). *The master and his emissary.* New Haven: Yale University Press.

McGuinness, D., & Symonds, J. (1977). Sex differences in choice behaviour: The object-person dimension. *Perception, 6*(6), 691–694.

Ingalhalikar, M., Smith, A., Parker, D., Satterthwaite, T. D., Elliott, M. A., Ruparel, K., et al. (2014, January 14). Sex differences in the structural connectome of the human brain. *PNAS, 111*(2), 823–828. https://doi.org/10.1073/pnas.1316909110.

7

The Beginnings of a New Motivational Theory, from the Engendered Brain

Paul says:

Where do we start? Amazon Books says there are more than 40,000 titles for 'motivational books', more than 4000 for 'motivational theory' and more than 1000 for just the word 'motivation'. Does the word even make any useful sense any longer?

What is it in any particular situation that drives human beings? is the fundamental question. What is it that focuses their attention and specifies their actions?

The pre-publication title of a new book caught our eye. *From Sabotage to Support* it said: *A new vision for feminist solidarity in the workplace.* The start of the advertising push says:

> Women are acculturated within systems that encourage them to sabotage one another; this book shows how they can break free of this cultural programming and use whatever privilege and power they have to raise each other up.

Joy Wiggins and Kami Anderson advocate that the only way women can successfully support each other is by addressing the varying intersections of our individual power.

We wonder about those kinds of over-arching statements coming, as they do, from a particular culture at a particular point in history but being generalised to all women everywhere. While this call to action might be useful for generating rage among those to whom it seems culturally appropriate, is it not the case that, worldwide, both men and women have different ways

© The Author(s) 2020
K. Lanz and P. Brown, *All the Brains in the Business*, The Neuroscience of Business,
https://doi.org/10.1007/978-3-030-22153-9_7

of expressing themselves that stem from their own particular cultures? And in any event proposing to grown-up women exactly *how* they should behave does not, in general, change very much. Or if sometimes it does, it leaves no room for that special quality of maturity—adaptive flexibility.

Let's go back to some basic biology from Chapter 2; and then to the new science of epigenetics that is beginning to work out how it is that the way we see things might have an influence on our behaviour. That's what the new motivational thinking starts from, not from an assumption that all individuals are essentially the same or that there is a political agenda to fulfil.

***

Women carry within themselves the possibility of creating, carrying and completing the bringing of new life into the world. Men stop at stage one of that process. For men, by the time that all stages are accomplished, it is nine months ago that whatever the experience of conception was for the man took place.

Not so for the woman involved. Life has been forming inside her and her body has been going through remarkable changes on the way. But most of all—and it is a fact so rarely mentioned that it is not part of common social discourse at all—she has put her life at risk. Despite all the advances in western medical care, all women risk their lives in child birth. There is no similar life experience for a man arising from the natural processes of reproducing ourselves.

Just to take in some figures that specify risk, in 2013 for every one-hundred-thousand live births in Sub-Saharan Africa, 550 mothers died from complications of the birth. In the US in 2011, it was just about eighteen. But despite the astronomical costs of healthcare in the United States, that was the highest rate in the developed world. In Australia, Denmark, Norway, Sweden and Switzerland it was more than three times better—five deaths per one-hundred-thousand live births.[1]

But evolution has been going on much longer than modern health care practices have been in existence. It is central to our contention that there are profound differences between men and women so that, within *her* biology, a woman *knows* that the gift she makes of herself in conceiving a child is a *total* gift of herself. She puts her life at risk. And her physiology is organised around the fact that she does not even have to be sexually interested or excited for that to be the case. A man has to reach a sexual climax of one

---

[1] https://www.weforum.org/agenda/2016/05/what-s-behind-america-s-shockingly-high-maternal-mortality-rate/.

intensity or another to be able to reproduce. Not so for a woman. For her a sexual climax is entirely separate physiologically from anything to do with reproduction.

So we want to start this exploration of searching within the applied neurosciences for a new motivational theory by having a starting point that makes the case for saying: The way men and women see their worlds is fundamentally different; and that difference comes from their basic reproductive biology.

What happens within the reproductive biology of both men and women is different behaviourally, too. It can be summed up in another aphorism that created itself when, in the early 1970s, I was running a clinical trial and training programme on the treatment of sexual difficulties, funded by the Department of Health. Sexually speaking,

Men want to want. Women want to be wanted.

The underlying reason for this comes from the physiological differences already described. A woman needs to know whether a man is 'worth it'. It's a difficult decision, and often proves to have been an error. But deeply embedded in the neural substrate of the woman's non-consciousness is, we think, a perception that *if* she is to put her life at risk *then* the man had better be worth it. So she needs to be in a position to be the decider about whether or not she will accept a sexual partner. It is she who says 'yes' or 'no'. The man can ask, the woman makes the critical decision.

Of course there are huge variations on this in all sorts of life situations: and plenty of patriarchal cultures where the woman's freedom to say 'no' is severely limited as is her economic freedom and independence to be other than strict social norms dictate. Such social differences, however, do not detract from the essential underlying biological point that is being made. Which is that at the very heart of our existence there is such a profound basic biological difference between men and women about what existence is about that it creates a profound difference between the way men and women see their world.

If we are on the right track in thinking about the essence of being female from this biological perspective, then the perception of men holds life-threatening possibilities for a woman but, if she is going to be a woman who creates new life, these strange beings called men are inescapably necessary. So they are life-giving as well as life-threatening. What a paradox to have to contain.

The same is not true for men with regard to women. The separation of the act of making love from its potential reproductive consequences is, for men, a vast separateness. There may be social consequences, but these are not fundamentally biologically built-in as is the case for a woman. Depending on the man's own maturity as well as his social context, he will deal with the living outcome of the act of making love as he will deal with it; from callous abandonment to total absorption and everything in between. Whatever his reputation, he has not put his life at risk.

We generally take all this bringing-of-babies-into-the-world so much for granted that it is rarely spoken of in these terms but therefore, alas, largely not in social awareness: and we know of no other book addressed to the world of women's development in organisations that approaches it at all. But once we start to take a biological view of behaviour we cannot conveniently skip bits about what we do know of our neuro-bio-physical make-up because they have never been brought into this kind of consciousness and context before.

There may be those who think that in any event they ought not to be. That social theorising and behavioural observation will provide adequate answers. Yet they lead to no firm agreed conclusions, just differences of opinion. And while that might be good for business school teaching, we want to get matters into working consciousness that might help our understanding of the curious matter that men and women *do* seem to be different; because we want to maximise on the advantages that women might bring to organisations once such ideas *are* in working consciousness.

***

The two aphorisms that have appeared in this book now form the basis of developing new thoughts about motivation—thoughts that have the power, we suggest, to lead to different ways of thinking about women in organisations. Both derive from the nature of the biological basis of the search for personal identity. The two aphorisms are:

> Men do what they do to find out who they are.
> Women do what they do because of who they are.

and

> Men want to want. Women want to be wanted.

In cultures and sub-cultures where there is no or very restricted opportunity for women to explore their own identity other than as defined by men, what we are now formulating will not apply at all for its rests upon the western liberal assumptions of the rights of the individual. But each of us having lived extensively in very different cultures from those of our origins, we would not be surprised to find that in cultures that do set out to define a woman's existence in patriarchal terms women have found their own ways of evidencing the feminine assumptions contained within these aphorisms, or long for the freedom to do so.

In Chapter 2 we showed that all behaviour has a neurobiology underpinning it. One very interesting consequence of this is to do with the way we see things.

Look for your best friend coming out of a crowded tube station. Your capacity to spot that friend among a sea of faces would be very much greater than if you were trying to spot a stranger who had been described to you. The accumulated experience that you had of your friend would incorporate all kinds of details that face- and movement-recognition technology is now replicating in algorithms. For you, though, being human and with feelings, anticipation would be hard at work and a great deal about your friend would be up in consciousness or close by for immediate recall.

Imagine now again a situation that has been described earlier in the book where there is no such anticipation. You are walking in your favourite high street when suddenly you see one of your oldest friends who moved out of the area four years ago to take a job abroad but you had rather lost contact. And here she suddenly is. The look on both your faces, the open arms, the tone of voice, everything that conveyed surprise and instant delight would have occurred without a moment's conscious awareness.

Yet no other person in the high street had elicited such reactions from you—perhaps quite the contrary. Everything you evidenced came not from what we are used to calling memory but from millions of patterned connections belonging in part particularly to that friend but also to lots of the other attachments of your life. The way you 'saw' your friend was the product of experience encoded in your neurochemistry.

Imagine the same high street and the same pending encounter with your friend, but on this occasion your life is suffused with great sadness. Whatever the cause, it is so. The spontaneity of your friend's greeting would not be matched with the same joyfulness as first described. The smile would be harder to find, the spontaneity less. Your friend might instantly spot your distress and sadness and, being a good friend and having known you well, respond instantly with much concern. And although all this seems to be conscious activity on both your parts, everything that is driving it is non-consciously created from within the complexities of your feeling system. That tells us two things and reinforces another.

The reinforced bit is to do with understanding that behind every single action or thought we ever have there is a biology at work. But what it additionally tells us is that *the way we see things internally and externally* determines how we act. At least, therefore, we have a scientific basis for beginning to re-consider motivational theory. And the science of it, called epigenetics, goes something like this.

The very structure, cellular composition and connectivity of our brain encodes our experience. It cannot encode anybody else's experience unless we have somehow made it our own. It is only through the lenses of that experience that we can make sense of the world—which is why going to a strange country is 'strange'. Until we have built up enough experience in a new context our brain does not have any reference points for dealing with the newness except to allocate it to a category called 'new and strange'.

'Appropriate' is entirely from within the individual responding to whatever is happening in the internal or external world. That is why we are emphasising an understanding of motivation that is individually, not generically or group, focused.

We now have the basis for thinking about motivation and the individual. The primary mechanism underpinning behaviour is the way an individual's emotional patterning has taken place: and that, in consequence, is what creates their perception. Their perception—the way they visually see and 'see' as in 'understand' anything and everything—is where the biological specifics of behaviour are generated from.

What, then, motivates any individual? The answer is very simple. It is their own internal, private, personal, emotional system. It has been developed like the interior of a house—unlike anyone else's, however much the outside may look the same.

This seems to create a huge problem for organisations, though. How is it possible to get people's energy flowing (thrive) towards the strategic and operational goals of the organisation if everyone's system is different? Never mind about differences between women and men. *Everybody's* different.

That's true—except for the fact that the basic system is essentially the same for men and the same for women though *somewhat* different between them.

Think of fifty cars all displayed at a motor show. Manufacturers want to emphasise the differences, and they do that in all sorts of ways. But the internal systems are essentially the same, with principles largely unaltered since Karl Benz created the first effective car in 1885. Some form of energy has to be fed to a reciprocating engine to give power. That power has to be controlled by being able to increase and decrease it at will. The vehicle has to be able to start and stop. And it has to be possible to steer it accurately.

It's much the same with people. The problem with motivational theory to date is that it has had no real understanding of the way (thrive) energy was generated inside the human being, and presumed that the stimulus to it could only come from outside the person. It was a manager's task to motivate workers. But our understanding from the modern applied neurosciences is that such a task is misconceived. A manager's essential function is to not only know what the direction of travel is but to create the conditions under which the self-regulating systems called people—driverless cars, in our simple analogy—direct themselves in the organisation's interest.

An important neuroscientific concept to introduce here is an understanding of the emotions of attachment. In our understanding of the eight basic emotions (see the Preface), surprise/startle is like starting the engine. Activating the escape/avoidance emotions (survive) is like trying to drive with the brake on. Immensely useful though brakes are, they are only useful in context. There are times when it is absolutely right to be fearful, angry, disgusted, shamed or sad. But knowing how to use the brakes properly and how to manage a handbrake is what makes for a smooth ride. Managing escape avoidance emotions properly is about regulating things well where danger would otherwise lie. All this is about to keep the power under effective control. In humans, the emotional energy and neurochemical responses that comes from the escape/avoidance emotions drives attention inwards, in a lightning fast response to look after the person first of all. Anger may make it look as if energy is going outwards, as in a fight. But it is only in the service of looking after the person who needs to fight.

What needs releasing in people is the brain and attentional power that's available in the car for the open road.

The means of doing that is to do with the attachment emotions, and the power is released through the quality of relationship within which any corporate activity takes place. If the part of the brain that manages the emotions, the amygdala, senses danger of any kind, it manages everything from a self-preservation perspective. But if it is certain there is no danger then it can permit the productive, outgoing attentional energy of the attachment emotions to flow (thrive)—just like having a car cruising apparently effortlessly in top gear.

So motivation is about leaders creating those conditions, with the complication that while both women and men have attachment emotions and both respond to the quality of the relationships that surround them, the triggers as to what releases the attachment emotions in women as against men seem—not surprisingly—to be different.

Let's go back to the aphorisms and the understanding that *if* women and men are fundamentally looking at their worlds from different perspectives, the triggers that motivate them would be different.

For men it's fairly simple. If men essentially know themselves by their achievements—what they *do*, not who they are—and they do that competitively, then they will respond to achievements-driven systems in which relative position (dominance) is also in play. Men define themselves and each other organisationally with where they stand in the hierarchy of men. It's why organisational titles are so important to men. It tells them not only who they are but where they stand.

Women are not especially interested in titles for their own sake. They are interested in what freedom a title gives to link everything together. If men's brains are focused on problems, and women's on solutions, then what motivates a woman to function at her best is the freedom to do that because (in giving birth) it is what she most naturally does. To create life women have to *be*. To create life men have to achieve—and get the immediate gratification of doing so. That's the origin of the differing motivational systems in men and women. And organisations have, not surprisingly, evolved to date with a focus on how to please men's inner reward systems. Women coming into organisations have accepted what was offered because it is (biologically) not their job to complain; for in the creating of life they get the bigger prize even if it is delayed and maybe not much fun at the start.

8

The Business Case for Valuing Brain Sex Difference

Kate says:

Embracing Gender Diversity at Work—The Numbers Stack Up

The business case for gender diversity is strongly evidenced in research. Women are gradually eroding workplace barriers that have traditionally prevented their progress. In addition, no self-respecting corporation would now question the fact that gender diversity is good for business performance. Indeed, most large organisations are working very consciously and with good faith to create gender diversity across all levels in business. Even if sometimes ambition is set somewhat low, between 25 and 40% (we could ask why not at the 50% level to truly represent the markets business serve from a sex ratio perspective) the direction of travel is very clear now. But we are arguing not for equality as the goal, but for *valuing the differences* in creating sustainable organisations for the twenty-first century. And we believe the route is through e-quality—creating conditions under which feminine energy emerges as a quality of real value into organisations.

The business case for women in management at all levels contends that companies that achieve diversity, and manage it well, attain better financial results, on average, than other companies. The business case evidence has been around for a while. In 2004, Catalyst used three measures to examine financial performance: return on sales (ROS), return on invested capital

© The Author(s) 2020
K. Lanz and P. Brown, *All the Brains in the Business*, The Neuroscience of Business,
https://doi.org/10.1007/978-3-030-22153-9_8

(ROIC) and return on equity (ROE). The Catalyst findings (Wagner 2011) include:

- Companies with the most women board directors (WBD) outperform those with the least on ROS by 16%.
- Companies with the most women board directors outperform those with the least on ROIC by 26%.
- Companies with sustained high representation of WBD, defined as those with three or more WBD in at least four of five years, significantly outperformed those with sustained low representation by 84% on ROS, by 60% on ROIC and by 46%on ROE.

We must, of course, be careful not to confuse cause and effect. It may be that the higher performing companies were doing all kinds of innovative things to drive performance of which the close involvement of women at senior levels was but one. Companies at their best are complex adaptive systems, so the presence of women may be evidence of that fact, not cause. However, we think it is unlikely that companies that have every appearance of being chaotic, maladaptive systems are unlikely to involve women as senior board level. Such organisations would be part of the male high-end spectrum that confuses action with direction and that demands only performance, not the quality of relationship that inspires performance.

Lord Davies was pleased with the results of the efforts of FTSE companies in generating better gender diversity at board level, since his first major analysis of Women on Boards in 2011. He concludes in his 2015 review (Davies Review 2015) that while progress has been made, there is still more to do.

> The business case is even stronger today as Chairs report on the positive impact women are having at the top table, the changing nature of the discussion, level of challenge and improved all round performance of the Board.

More recent 2018 McKinsey research (Hunt et al. 2018) into delivering growth through diversity in the workplace reported that 'Gender diversity is correlated with both profitability and value creation'.

The McKinsey research demonstrated a positive correlation between gender diversity on executive teams and key measures of financial performance such as earnings before interest and tax (EBIT). Their study also revealed that companies with gender-diverse senior teams were also more likely to maintain longer-term value creation than the companies without a critical mass of women at executive levels.

In short, businesses that are competent at harnessing the best of what women have to bring by welcoming and valuing the difference at executive levels and at sufficient scale, perform sustainably better than companies that fail to do this. Wide-ranging and long-standing research has consistently shown this to be a modern business truth. We do nevertheless have to consider the possibility that the women who have achieved Board positions are more male-brain dominated than many of their female counterparts. If that were the case it still doesn't undermine the proposition that something about having more women in senior positions seems to have a very positive impact on performance. Science will doubtless go on elucidating the reasons why.

Glaciers Do Move—But Slowly

While there has been progress in the direction of gender diversity on British boards, it is glacially slow. It is nowhere near gender balance. Commissioner at the Equality and Human Rights Commission, Stephen Alambritis, said, after the 2011 Davies report:

> At the current rate of change it will take 73 years for women to achieve equal representation on the boards of FTSE 100 companies. We need to speed up progress. This is not just a moral issue. Our businesses are paying a penalty; there is evidence that more diverse boards take better and more responsible decisions. (Women on Boards)

At last year's World Economic Forum (WEF) in Davos Saadia Zahidi, the WEFs head of social and economic agendas observed: 'The overall picture is that gender equality has stalled. The future of our labour market may not be as equal as the trajectory we thought we were on'. WEF estimated in 2017 that at current rates it will take 217 years to close the gender pay gap and 107 years until there are as many female politicians as male (Neate 2018).

Though simple number counting is still the current frame of reference it is not the full answer by any means. It creates easily-understood data, of course, but women's contributions at senior management levels are much more subtle than just their presence—though presence *is* one of the effects that, in the absence of women, has no chance to have its effect. At this present state of knowledge, presence is what is being counted.

These efforts by governments, business and organisations such as the WEF are a force for good and must continue. However, it is our strongly held belief that there is an accelerating force already sitting inside each of the businesses we work in, that can catalyse access to the brain diversity that is already in the building. Access to this accelerating force is through active, powerful discrimination, as referenced in Chapter 6, through opening up the eyes of businesses to what they are missing, corporately, that allows brain gender already in the building to be heard. Cultures and ways of working that are advertly or inadvertently encouraging female brains to have to show up like male brains are losing access to a potent source of business performance.

Creating Faster Flow—Powerful Discrimination

The more potent angle to the discussion and the key argument that we set out to make in this book is *how the difference that brain gender diversity brings is actually being leveraged* inside the organisation once it arrives in the building.

As coaches to senior women in business globally, we are continuously reminded that it is still very much the case that what is seen as, measured and called out as successful behaviour in the workplace largely demands alpha male qualities. Most large organisations have been created over the decades by men for men. This shows up in all sorts of work practices such as the qualities that get looked for in recruitment, the way that power gets noticed measured and bestowed, the way meetings are run, the way feedback is given, the performance review systems and so forth—all of the practices that combine together to form the culture of the place called work.

In order to get ahead within certain corporate cultures, it is still the case that many women in the workplace are expected to be the best men they can be. And many men who have a more female brain are expected to behave in more alpha male ways. This does not create the ideal conditions for the brains of these women and men to be in *thrive*. Being good corporate citizens, they will do their best, of course. But they are not able to *give* the best of themselves. A good deal of energy goes into being what is required, not maximising on who they are. That's the loss to organisations and the point of leverage.

Please do not misunderstand, alpha male behaviour can be very positive, powerful and productive. There is absolutely no doubt about that. It can also be very destructive. If male alpha behaviours become the dominant way of going on, in terms of work culture and work practices, an environment becomes created where the best of the more female brained individuals is either simply not fully accessed or is actively shut down. One point of view has it that the last financial crisis was caused by alpha male behaviours (Coates 2012).

The best of organisations set out to harness the best of all the brains in the business: the best of the most male brain all the way through to the best of the most female in the building makes not only for better performance overall; it creates the conditions under which people are free to willingly give of their best. That is a priceless competitive advantage. It is also the basis of the new motivational theory. The organisation can require people to come to work, but it cannot command how they manage their brains. It can however create the conditions under which the brain gives of its best rather than reserves much of its resources to struggle to be there at all.

Separating out alpha male positive behaviours from alpha male negative behaviours can be very productive too. Sometimes the dividing line is small, but it is nonetheless very important. Let's take a quick look.

The Fine Line—Alpha Positive Meets Alpha Negative

Alpha positive behaviour is not aggressive. Alpha positive will sit back and watch everyone get on with their activities but will stand up and fight for what he believes is right when the time comes. This can be highly productive, creating focus, pace and action at the right time. Alpha negative behaviour will want to win almost every round and will dominate others to do so. The risk then is that too much internal competition is generated (male testosterone flows) and the focus on winning in the market place reduced. It is also unlikely that the alpha male who wants to win every time is always right. He becomes (in his own estimation) *the* brain in the business but huge additional power gets blocked.

Alpha males are leaders and don't wait to be led. An alpha male generally assumes the leadership and does not wait to be crowned as one. This can clearly mean that alternative more collaborative leadership approaches are not recognised or allowed any currency. Women take their power differently than men on the whole, as do non-alpha males. Alphas don't like waiting for instructions from anyone. They look around, judge the situation and make a decision on what is required to be done. Alpha males don't shy away from making tough decisions just because they are afraid of failure. An alpha male displays the capacity to solve every problem. They look after their comrades and give everyone around them a sense that they are in safe hands all the time. This can create a positive sense of belonging and enable people to do good work. But if it is unrelenting, overdone, it can create the environment for passivity and 'group think'.[1] Another alpha negative downside is that it can feel over-controlling. The way a more female brain creates a sense of productive energy is different—it is more through tuning into people's emotions so they feel heard and understood, and that they feel a part of something that is working well.

Alpha males speak less but pack a punch when they do. This can be incisive and cut to the chase in a business environment in a positive way. It can also intimidate and close other opinions off without the alpha being aware that good contributions are being shut down. There is often an assumption that people need to have the guts to speak up and little respect for those who don't. This is the tip into the alpha negative and where good opinions go missing.

Alpha male personalities are likely to create a stir in a room by their mere entry or presence. Whether it is because of impeccable dressing, a firm handshake or effortless body language, alpha males tend to naturally command respect and attention. Depending on the situation this type of impact can be very important for business performance. Deal making, handling certain types of conflict or leading a group of like-minded people would all be examples of where this aspect of alpha positive can be a useful approach. The forceful presence, dominating the space, is not how more female brained individuals like to take their authority.

Case Study: An Alpha Male Interview Misadventure

Miss B said she applied for the job because she thought that "the job sounded just right for me, it listed a lot of perks and sounded right up my street."

After getting through the first round she went back for a second interview on Monday in front of the CEO.

"I was really excited to meet the CEO, I'd watched all his talks online, but from the moment I walked in he was disinterested," she said. "He didn't stand up to greet me, he was just on his phone."

Miss B said the interview started in an "utterly bizarre" fashion in which the CEO picked on her music tastes before revealing he was scrolling through her music account while they talked.

"He started asking personal questions, about my childhood and if my parents were still together," she claimed. "I thought it was really bizarre for an interview."

---

[1]Noun: **group-think**—the practice of thinking or making decisions as a group, resulting typically in unchallenged, poor-quality decision-making.

Miss B claims that the CEO then began "tearing apart, line by line" everything she had submitted in the written part of the application process.

"He completely ruined my self-confidence, he made me feel worthless, like I had no talent," she said. I was very close to tears at several points in the interview.

He called me an under-achiever, told me I knew nothing, then asked me how I thought the interview had gone.

I said I didn't know, and then he told me I didn't know anything, and everything that I did was wrong—my body language, how I spoke.

I felt he was attacking my character. He told me he could tell I was upset and yet he continued to push.

"I've been in this position before," she wrote in response to the job offer. They tear you down, abuse you, take you to breaking point and then they take you out to dinner or buy you a present to apologise and make it seem like they're the nice guy.

"This job is supposed to be the present. I don't want it."

Top Talent: Finding It, Keeping It—The Biggest Kicker Yet the Biggest Challenge

In my research, I am helping clients to pinpoint the lost brain potential in business-critical populations—such as key talent. Key talent are people who are being specifically targeted for development, promotion and, importantly, retention with a view to becoming the very top leaders of the future.

Top talent can be up to eight times more productive than an average performer. A 2012 study by McKinsey revealed that for jobs involving high levels of complexity, high performers are an astounding 800% more effective than an average performer (Keller and Meaney 2017). This is a significant productivity kicker. McKinsey coined the expression 'the war for talent' (Chambers et al. 1998). For Gen Y and Gen Z, retaining top people becomes an ever-increasing challenge. In the Conference Board's 2016 survey of global CEOs, 'Failure to attract and retain top talent' was the number one issue ahead of growth and competitive intensity (Conference Board 2016).

Losing talent is an expensive business. In a 2018 study, a team from Deloitte established that the total cost of employee turnover may range from tens of thousands of dollars to at least one-and-a-half to two-times an individual's annual salary (Bonnie 2017). Forbes stated that for entry of highly skilled-level employees, the associated cost comes to 400% of their annual salary, respectively. In US companies, employee turnover already costs $160 billion a year (Borysenko 2015). These are startling statistics.

Triggering Survive Is Bad for Business!

In some of my studies into untapped brain potential in key talent popula-tions, my research is revealing that up to 30% of the executives are spending much of their day with their brains more in a **survive** than a **thrive** state. We know from the neuroscience that the *prefrontal* cortex will be doing sub-optimal work when the limbic system signals a survive response. This is the way evolution has designed us. Having 30% of the brains belonging to your key talent in survive mode much of the workday is not optimising your business performance. Identifying what is causing a survive response and changing it will significantly enhance the performance of the talented brains that an organisation already has in the business.

The survive triggers will occur at three levels. These are firstly the individ-ual level. Our own Trustprint™ determines this and this is deeply individual. It is helpful to reflect upon this personally and have good line of sight on one's own survive triggers and habituate effective ways of managing them. I have developed an app to support leaders to do this. Good leaders will notice when they are triggering survive in team members around them and will flex their style to avoid this in the knowledge that they cannot access the best of the other person's brain if the other is in survive.

The second level is systemic. What is going on at an organisational level that might be causing survive? One of the common systemic issues we come across in our work is that many business cultures have become too trans-actional. They are measuring performance KPIs to such an extent that they leave no time for people to form real relationships whereby humans can tune into each other to a sufficient level to understand and support one another. Days at work end up being packed with meetings, agendas are dived into at such a pace that connecting with all the brains and the people they belong to does not happen—all in the service of so-called efficiency. From a brain-based point of view this does not work. People who do not feel connected, seen, heard and somehow validated cannot neurobiologically perform at their best. The worst-case scenario is that key talent experiencing triggers that cause that level of survive in their work environment are more likely to leave an organisation. We have seen above the cost to the business of losing this type of hire. In some of my research, some of the brightest brains in the business, which were also slightly counter-cultural and thought a little differ-ently, were the brains that felt the least welcome and were therefore the ones that were more likely to leave. One of the owners of such a brain once said to me, 'Kate I am going to leave this place because my face does not quite fit. No one will really notice – and no one will really care. They will replace me with someone a little less expensive'. Wow! I thought—how sad to feel like that about coming into work most days. And of all the brains in the group I was researching at the time, his was the brain I would least like to leave the business were it my company.

The third level is the macro environment much of which one cannot change and has to accept is outside the sphere of influence. So—issues such as aspects of Brexit are outside an individual's or an organisation's control. There will always be environmental factors that cause an appropriate survive response. Intelligent business leaders do not want to add to these inadvertently by internal factors. They also recognise the external realities and set about turning them into the best competitive advantage that they can.

Which Brain Types Are Not Optimising Thrive?

It's fascinating, but perhaps not surprising, that my bespoke company research has also demonstrated that the brains spending 30% of their working day in survive are the brains on the more female side of the brain-sex spectrum. The more male brains are tending to show much higher thrive scores and are tending to enjoy the work practices and have less of an issue with the work cultures we have been investigating. It's not surprising because not only have most workplace cultures been created by men for men but men are amazingly blind to the differences that the female brain brings to corporate advantage. And why should they not be? It is a part of almost every man's experience that he was looked after by a woman whose primary task was to look after him. Why should it be different in corporate life? The answer is that women are in the corporate life on their own terms, exploring their own lives, not wanting endlessly to serve men.

The causes that trigger survive are often subtle and go deep into the essence of the difference between what it is to be a woman in a workplace that is subtly designed to favour a male brain. And let's not forget what it means for a more female-brained man. These brains too showed lower thrive scores. So—though a man with a more female brain will still have different ways of going about tasks and by nature have higher testosterone levels than the average woman, accessing the best of brain sex difference is very nuanced activity and one that high-performance modern leaders take very seriously.

Creating a deep understanding that means these subtle differences are appreciated and then establishing a working culture and work practices that acknowledge, accept and enable the brain-sex difference, is a huge source of latent productivity for any business. A brain that feels its owner is seen, understood and validated and that not only has it been asked to the party, but it is also being invited to dance is a brain that will really perform and thrive. Research shows that work cultures that are in the top quartile on trust are up to 50% more productive than their peer groups (Zak 2017).

Innovation and Agility—The Polymath Brain for the Future

One of the biggest challenges for modern businesses, especially in terms of remaining competitive in the face of the focused might of China, is creating innovative solutions fast and then being able to execute with agility. Many businesses are finding that their structure and processes are simply not fit for this purpose today. We are increasingly working with teams where the challenge is how to form more fluid, connected working structures such that a multitude of perspectives are brought to bear, heard and integrated quickly and acted upon fast. This is often a question of blurring the boundaries of the disciplines involved in problem-solving and having communication that is very fast and not hierarchically driven and where trust underpins the capacity to sort out differences quickly. Protecting one's turf and reinforcing a silo mentality simply does not cut it these days. Yet there is still a lot of it about. Diversity as we have seen drives better business results but accessing what brain diversity has to offer is difficult to do. Diversity per se is not the answer. It is **how** the diversity is allowed to join in that makes the performance impact. Accessing brain diversity effectively is the aim of this book. The following case study is an example of how to create an environment where becoming a polymath and operating with a polymath mindset enables fast, fluid learning and execution.

Case Study—A Global Institute

A top international university set up a global institute with a view to developing polymaths, people able to think critically yet broadly, able to know enough within a relevant discipline to interrogate and access it but without needing to be a deep subject matter expert and able to work across disciplines to solve big issues. The founding premise of the Institute is that the types of problem that we need to have solutions for as a human species are too big and too broad to be solved by specialist thinking. In the modern world, we must work together in a far more connected, integrated way. Physicists, biologists, chemists, engineers, finance and business and people specialists need to be able to think together, communicate together and be curious together about the big problems the human race faces, and to act quickly to go from idea to prototype and beyond.

The research projects tackled by the Institute train polymaths. And importantly the whole set up and ways of working and culture of the Institute reflect this integrated, fluid, non-hierarchical, intergenerational way of being. The essence of the way the Institute works creates thrive in the minds of the people who choose to join it. There is significant gender diversity starting with the two co-leads—both scientists—a man and a woman. There is significant ethnic and LGBTY diversity and what is prized above all else is the capacity to come in with an open mind and learn together and co-generate solutions to complex problems—such as new energy sources, and antimicrobial surfaces to name but two examples.

The Institute is not bound by departmental boundaries. It exists across all elements of the university. It faces into industry, actively working with large companies to help them think about, model and test new ideas that might help them solve some of the big challenges they face (such as renewable energy). The teams of people in the Institute are fluid, fast-moving, open-minded and respectful.

The people and their ways of working practise what they preach, and it is a vibrant, successful thrive environment that attracts bright young people who want to be part of it. This type of approach could translate effectively into many corporate environments if they chose to think a little differently about how the task gets done.

Summary

As a human race what we do not have much of is—time. The rate at which we are destroying the natural resources that the planet provides us with mean that we can no longer count upon the survival of the natural world as we know it to support the human species. Nature of course will go on, long after the human race has become extinct. As a global society we need to work differently, think differently and live differently. Living sustainably within the resources that we consistently replenish and nurture is a must. We will have to access the best of all the brains in our businesses to find ways forward that work. Truly enabling the best of the female and male brain to work together, across generations and out-with hierarchies, intelligently, freely, generously and quickly is an imperative. Big business has a huge role in releasing the creative power that already sits within it. Building truly brain-friendly environments based on valuing brain gender difference is becoming a real possibility.

References

Bonnie, E. (2017, April 17). *Employee retention: The true cost of losing your best talent by Emily Bonnie*. Wrike.

Borysenko, K. (2015, April 22). *Talent management and HR. What was management thinking? The high cost of employee turnover*.

Chambers, E. G., Foulon, M., Handfield-Jones, H., Hankin, S. M., & Michaels, E. G. (1998). The war for talent. *The McKinsey Quarterly, 3,* 44–57.

Coates, J. (2012). *The hour between dog and wolf: Risk taking, gut feelings and the biology of boom and bust*. London: HarperCollins.

Conference Board. (2016). https://www.conference-board.org/webcasts/ondemand/webcastdetail.cfm?WebcastID=3587. 9 February 2015.

Davies Review. (2015). *Women on Boards Davies review five year summary October 2015*.

Hunt, V., Prince, S., Dixon-Fyle, S., & Yee, L. (2018, January). *Delivering through diversity*.

Keller, S., & Meaney, M. (2017, November). *Attracting and retaining the right talent*. McKinsey and Company. https://www.mckinsey.com/business-functions/organization/our-insights/attracting-and-retaining-the-right-tale. Accessed September 2018.

Neate, R. (2018). Wealth correspondent. *The Guardian*, Rupert Neate *Wealth correspondent* @RupertNeate Tue 18 Dec 2018 18.11 GMT.

Wagner, H. M. (2011). *The bottom line: Corporate performance and women's representation on boards 2004–2008*. Catalyst. Accessed 31 March 2019.

Women on Boards. https://www.gov.uk/government/news/women-on-boards. Accessed September 2018.

Zak, P. J. (2017, January–February). The neuroscience of trust. *Harvard Business Review, 95,* 84–90.

9

How to Ignite All the Brains in the Business

Kate says:

All the Brains at Work

At the applied neuroscience workshops we run with the senior business executives, we get two predominant responses. The first is: 'Aha—that explains it!' The second is: 'So what do I do about it? How do I use this information in practice?'

So here are some real business examples of the 'How to's' because actually that is what really counts. If leaders want to access the best of all the brains in the business then it is the day-in, day-out work practices and the workplace culture they create that is mission critical to success.

Through the course of my research with busy executives, I have been developing and testing easy-to-remember, brain-friendly models with busy executives that they can deploy in a wide variety of situations.

Trust

Trust is one of the eight primary emotions. It sits at the heart of thrive. Brains that feel trust are brains that are capable of learning, are creative and resilient. When the brain is in thrive it will be producing the right amounts of oxytocin, dopamine and serotonin for optimal learning and optimal performance

© The Author(s) 2020
K. Lanz and P. Brown, *All the Brains in the Business*, The Neuroscience of Business,
https://doi.org/10.1007/978-3-030-22153-9_9

to occur. That powerful guardhouse of the brain, the amygdala, reacts to make us the most adaptive we can be (in order to ensure our survival under all circumstances) which is three times more quickly than we can humanly begin to respond cognitively to any situation. So ***feeling trust first***, before we start to think, is the only way that we can get the best of our brains.

Trusting someone at work does not mean that you actually have to be friends with the person in question or necessarily even like them. We end up working with all sorts of people, many of whom we would **not** choose as friends. So there are two key elements of trust that are important in the workplace.

The first is the feeling that this person has 'got your back' and would not take any action or say anything that might harm you or your position at work. This type of trust signals that the person is part of your 'in-group' in the workplace (Eagleman et al. 2018). Brains recognise and respond pre-consciously to 'in-group' signals within 85 milliseconds. Where this type of trust reaches a deep level, you will be prepared to share some of your own vulnerabilities (such as a mistake you made at work) with them and trust that they will support you rather than use this information against you or think less of you.

The second element of trust is the trust that the other person is competent to do the job in hand, that they have the necessary knowledge about the actual content of the tasks in hand. This competence relates to both the technical side of the role—the hard skills—and the emotional/relational side of the role—what used to be called the soft skills but are now coming into the domain of serious science. Clearly, some people are better at the technical side than the relational and vice versa. As we have seen throughout this book, there are female–male differences that are more common in one sex than the other despite our mosaic brains. A team that has very high trust among team members (because team members feel emotionally and psychologically empowered with their team colleagues) is able to access the best of what each team member has to offer very quickly. This means that teams that trust can call upon whom they need with the hard skills to sort something task-related out and can equally well call upon team members with soft skills to bring their emotional intelligence—without hesitation or fear. Many team members will have both types of skill to greater and lesser extents. High team trust enables colleagues to bring the best of what they contribute to the team's efforts. This is evidence of intelligent emotions in action.

Trust is like an elastic band. You can stretch it during which it becomes thinner and less resilient and **mistrust** starts to build. Just like an elastic band, under enough stress it will snap at some point. If it snaps into **distrust**—it's game over. Once we actively distrust someone it is very difficult indeed to work with them and certainly optimal brain conditions cannot be achieved. It is also very difficult to re-establish, and needs serious willingness on both sides to do so which, once trust has been lost, is likely to be in short supply.

It takes serious effort to rebuild trust once it has been broken in this way. This takes time and a series of positive, practical experiences with the person in question to rebuild any level of trust *but* it will never be quite the same again. The amygdala in our limbic system plays an important role in encoding for memory. It does this based on the emotions that it experienced at the time the memory got laid down. The emotion underpins the meaning the brain gives to a particular situation. The brain will lay down memory that reminds us to avoid or be very cautious about a situation or person that triggered survive emotions in us or, equally importantly, created the sense of loss and a change in our world that broken trust engenders.

If someone has broken our trust (especially by not 'having our back') then the limbic system will be on red alert around that person for a long time and we are unlikely to get back to 100% trust ever again. There will always be some degree of energy that is diverted from optimal performance, to managing an element of distrust in our emotional limbic system.

Like a bucket of water that has been filled one drop at a time, a major act that results in distrust is like forcing a hole in the side of the bucket at the bottom. It leaves the water leaking out and the bucket empty. The slow, painstaking process of refilling the bucket one drop at a time has to start all over again, yet the hole is not easily repaired and takes effort, will and understanding from all sides, so that the bucket can gradually be refilled without leakage. It's a difficult job. That's why regenerating trust is so difficult.

The RICH™ Communication Model

The one activity that is always present at the heart of achieving effective sustainable, trusting relationships is communication. Each and every human interaction that is not working as well as it could requires improved communication to help create the understanding, the context for improvement, possible solutions and a mutually sustainable way forward. An act of communication that does each of these things will build trust. Building trust in every single act of communication is critical for creating the conditions for optimal brain performance.

As we have discussed, our brains have evolved over millions of years and, for good evolutionary reasons, they work bottom-up: through the emotional system into the cognitive, not the other way round. Effective communication requires, therefore, that we work in a brain-friendly way. That requires communication from the bottom-up of the brain.

The RICH™ Communication Model has evolved as part of my work in applied neuroscience in organisations.[1]

Given that, by nature, our limbic system is in a permanent state of alert, whether we mean to be or not, the first act in communication is to soothe and settle the limbic system of the person or people with whom we are about to communicate. It is a rule of thumb that I assume there is ***always far more emotional anxiety*** in a person or situation than one would ever imagine. If you go into a communication holding this in mind, it will always stand you in good stead. So, let's have a look at how RICH™ brain-friendly communication works (Fig. 9.1).

R for Recognition

The limbic system controls where energy will flow in the brain. If the limbic systems deems that you are safe, it will stand down and allow the powerful, cognitive brain, the prefrontal cortex, to fully engage. If the limbic system feels unsettled or downright scared then you will simply *not* have the other person's proper attention, so communication will be either useless or downright destructive. Thus, the very first thing is to ensure the other party's limbic systems is soothed and settled.

The way to do this is to start with what there is to recognise, appreciate and acknowledge about the other person's position—notably their emotional position on the matter in hand. Name very clearly and specifically what you can see about the other person's actions in the situation in a positive way (even if you disagree with them). It is important to validate them and allow them to *feel* seen.

---

[1]And a deep acknowledgement to the wonderful work of psychologist Virginia Satir.

RICH Communication Model

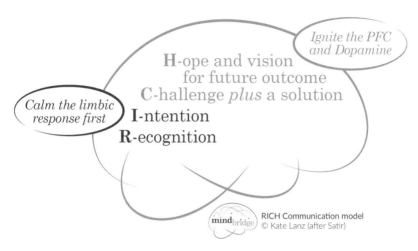

Fig. 9.1   RICH™ Communication Model (© Kate Lanz—after Satir)

Example—RICH™ Communication Model in Action

Let's imagine someone—Linda—has rushed into a situation at work without proper consultation, leaving you feeling on the back foot and out of the loop. Linda has just sent an email round copying in what seems like most of the department and you have not been part of any conversations but believe very strongly you should have been. The **Recognition** part of the conversation might go something like:

> Linda – I really appreciate your bias for action and capacity to just get going fast with an issue. You move a big agenda forward much of the time and that's great for the team.…

I for Intention

In its vigilance, the limbic system is very concerned about being caught out by something it was not expecting. So the next brain-soothing action is to reassure it that this will not happen by carefully indicating what this communication is going to be about. This way, the risk of having a nasty surprise is reduced *but* it is only reduced, not eradicated. There will remain an amount of wary watchfulness in the other's limbic system which is rarely entirely off duty. However, a heartfelt reassurance about the topic for discussion soothes the limbic reaction and allows the prefrontal cortex to start to come into play for the exchange that is about to follow:

Example—RICH™ Communication Model in Action

So—going back to Linda:

> In respect of the recent email on Project X, I would like to discuss with you how you and I need to connect early on this since it affects the whole team, and my role in particular, so it would be great to talk about that now – is that OK?

This is flagging to Linda the topic area to be discussed and specifically how it relates to you. It is also asking her permission to engage in the subject so she feels involved in having created the territory for this conversation to occur. She is partnering with you as opposed to being 'done to'.

The *recognition* and *intention* setting will go a long way to soothing both your limbic system and Linda's. This is vital, without this, a proper communication cannot occur.

Challenge Plus Solution Generation

The C of RICH™ is the substance of the conversation. Our PFC *loves* to solution. That is what it has evolved for—to make sense of the world, to imagine, to generate new ideas and connections. So setting out the challenge or problem you are facing with the person and beginning to create the context for solutions gets the PFC excited and involved. It is important to do this in a genuine spirit of wanting to connect and relate more positively to the person in question. If the underlying emotion is anger and wanting some kind of revenge for the annoyance caused, the person will feel that immediately. The brains of we mammals pick up the emotional signals way before the intellectual ones, though most of that is happening well below the threshold of immediate consciousness.

The idea is to create the context where you and the other person can generate solutions to the challenge or problem together. It is important not to present the solution as the only one but to give a clear sense of opening up the conversational space between you for a genuine exchange. People will *feel* our underlying intention in the first instance and respond to that rather than to what we say.

Example—RICH™ Communication Model in Action

Given the nature of the task in relation to Project X you and I Linda would benefit from having a proper exchange of ideas before certain other team members are involved in the debate. I would like us to explore and agree how we can partner more closely so that a joined-up approach from us can benefit the programme. What thoughts do you have about how that could work?

At this point it may be that you have to loop back around the R and the I of RICH in the event that the other person becomes defensive in order to resettle their limbic system (and indeed your own, as it will trigger too if theirs does) The art is to stay very tuned into the verbal and emotional signals and body language that you are noticing so that you can keep soothing the limbic response as you need to before re-engaging the cortex.

H for Hope

When we can see in our mind's eye and feel a hopeful, exciting picture of the future the reward centres in our brains become engaged. Dopamine is the key neurotransmitter for excitement, joy and reward. This gives us a buzz and a comforting, warm feeling. So verbally painting the picture from the future about how the situation will feel, sound and look when everything is working well will engage the thrive response in the other person. Of course, it has to be genuinely meant, felt and expressed. Anything remotely phony will be detected in an instant.

Example—RICH™ Communication Model in Action

Linda – you and I have a really pivotal role together as part of the success of the project. As we work even more closely together, we will have a very positive impact on the pace and quality of the outcomes. I'm looking forward to that. Thank you for this productive exchange. When shall we get together to talk about XYZ…?

The key to this whole process is writing and speaking with a simultaneous awareness of how it would be to be on the receiving end of what you are writing or saying—not as a manipulative process, but as a genuine act of wanting to create the best possible outcome. In such circumstances, the sense of genuineness that is created gets the two brains working together, not maintaining a position of opposition.

In a Nutshell

As you develop the skills of RICH™ communication you will notice the difference in flow and effectiveness. Importantly, you will feel the quality of your underlying relationships improving as a natural by-product of this brain-friendly way of connecting.

For some of you, maybe especially the ones with very high male brain sex scores, this might feel long-winded and convoluted. And so it may be, at first. Remember you are aiming to create the emotional response in the other on purpose, by design, not inadvertently. We are emotional three times faster than we are cognitively aware and rational (Wilensky et al. 2000). That is the way we have evolved. Limbic filtering is *always* present and we now know its effects are very much greater than was ever previously imagined. As with any high-performance, intricate system, it needs understanding and managing well for high-quality communication to occur.

The actual gender, what you have deduced about your interlocutor's brain gender and patterns of survive and thrive triggers will all be vital pieces of information for *how* you frame your RICH™ communication. For someone with a more female brain you would construct your RICH™ communication very differently from someone with a more male brain. As a test of your own application of the model, pick two colleagues of yours, one with a very male brain and one with a very female brain and imagine a RICH™ conversation that you might have with them and how you would frame it differently.

When to Use the RICH™ Communication Model

The RICH™ Communication Model is particularly helpful in a situation when there is a difference of opinion. It can also be used in the following situations:

   To structure a coaching conversation
   To provide constructive feedback when something is not working

To frame a town hall-style presentation

It is also really helpful as a parent finding a way to connect with a teenager when they are upset or misbehaving, we have found!

Meetings and the Brain

A 2017 study quoted in the Harvard Business Review demonstrated that, on average, leaders are spending 23 hours per week in meetings (Perlow et al. 2017; Rogelberg et al. 2007). This is 10 hours more than in the 1960s. Of course, meetings are very important for collaboration and bringing diverse brains together to problem solve. However, my research is showing that meetings are missing a productivity upside of between 30 and 40% by not actively harnessing the best of all the brains in the meeting.

Why is this? The main culprit is a combination of overarching company culture blended with meeting format. In my experience (and it's long and varied), it's rare that meetings actively seek out the opinions of *all* the people in the meeting. They want to get to solutions too fast. Secondly, it is rare that there is proper time and space given for people to express their full opinion and be deeply listened to as they do this. So meetings are too often focused on speed and getting through an agenda. A transactional efficiency is often the tone of a meeting rather than an open space where minds can meet and reflect together. *Minding* about meetings gets individuals to bring and give of their best, not just favour and focus on those who proactively speak up and take the floor while others who might be more introverted, or reflectively thoughtful in context, are mostly not supported to be heard.

The fact is that business still gets done and the action moves forward *but* there is a hidden loss of productivity through not accessing the neuro-diversity in the room. And as we have been discovering, different brains need different approaches to enable them to be in thrive and show up to their best.

The 4Cs Meetings Model™

During the research and focusing on brain gender difference, I developed and tested the 4Cs Meetings Model™: the aim being to access the best of *all* the brains in the meeting. The underlying principle starts from the knowledge that the brain works bottom-up. So let's begin with:

C for Connect

The limbic system settles when we feel safe, it knows there is no danger and works out (in nanoseconds) (Gawande 2011) that its owner is being valued and not judged (LeDoux and Pine 2016). So connecting with all the brains in a meeting before the substance of the meeting starts is really important in order for participants to be fully present and ready to contribute. This act of checking in and tuning into how people are feeling coming into the meeting may sound time-consuming but in my experience is the best investment of time for the whole meeting.

Case Study

A trauma surgeon in the States, Atul Gawande, discovered that when whoever was on duty for the trauma team rushed to the operating theatre from wherever they were in the hospital as the call bells clanged, if the team members spent a few precious seconds introducing themselves by name, patients stayed in hospital one day less than when operations were conducted by a team where the individuals knew each other only by function. And if there was a crisis during surgery, as is not uncommon, a team that knew its members by name seemed to get out of the crisis better than a team that had not introduced itself. A moment's connection became momentous in this way.

The connecting phase will vary in timing and style depending on who the meeting participants are and the purpose of the meeting. So for instance in a coaching supervision group where the work we are about to do is deeply reflective, emotional and personal, the check in process reflects this and we tune in carefully to what each person in a supervision group is *feeling* as they enter the supervision space. This may take a good half an hour of a three-hour session.

In a business meeting designed to make a single decision on one topic, the connecting process will be faster and needs to match the tone of the topic, as well as take into account the types of brain in the room.

For Example: A 4Cs Connect Technique

Share one word that reflects how you are feeling coming into this meeting and a short sentence about what you want to get out of the meeting.

A client came into the meeting saying she was tired through having been up twice in the night with her toddler but excited about the meeting topic. It was an important one for her project. Her desired outcome was a sense of alignment from the team on next steps, leaving no concerns under the carpet.

This method gets everyone's voice into the room early and quickly while acknowledging what is going for people as human beings with a life outside work. It also focuses the PFC on the solution and paints the picture of hope for what success might look like. So it both soothes the limbic system, allowing it to relax, and tickles the cortex in a positive way.

C for Compassion

Humans cannot help judging other humans. The brain assesses in an instant whether it considers another to be part of the in-group or the out-group. Some fascinating research by David Eagleman (Eagleman et al. 2018) shows that purely on the strength of a random label we start to judge others as out-group.

So we are incapable of switching off our non-conscious bias since, by definition, we cannot access the non-conscious quickly enough. Because of this we have to do something else instead. This is to recognise and articulate to ourselves as best we can what our judgements of others are and notice them. Then, in the instant, to quite deliberately work beyond them and commit to including everyone in the meeting in a way that is compassionate and judgment free *if* what we want is all the brains functioning at their best.

Because we are mammals we can *feel* when someone is judging us and listening to us through the filter of their opinion. And we can *feel* when we are being truly listened to and another person truly 'sees' us. The difference in the neurochemical response of the brain and body of someone who feels judged (survive response—cortisol and adrenaline) and the brain and body of someone who feels validated and valued (thrive—oxytocin, serotonin, dopamine) shuts down or creates learning and the capacity to contribute.

So deliberately engaging with all the brains in the meeting with compassion and an open mind will create the neurochemistry of performance. That's good for business.

C for Curiosity

In one of the research groups I was running with a large corporate client, one of the smartest brains in the group said to me 'The thing is, Kate, my face does not quite fit around here. I am counter cultural. When I am in a meeting no one seems to care what I think and unless I have a very specific point to make, I often dare not speak up – the vibe is too political. So I don't'.

The person who said this had one of the smartest brains in the group I was researching. Were that business mine, I would definitely want to know what his brain thought. He was a wonderful lateral thinker. Sadly, he was rarely asked what he thought and because of the constant feeling that he did not quite 'fit in' he largely kept his views to himself and spent most meetings in survive.

Curiosity is king—or queen. Ask. Ask each brain in the room what their opinion is on a matter. Be curious. If the person is in the meeting there must be a point in them being there—must there not? If not, then your meeting structures really do need refreshing! In the spirit of genuine curiosity, allow each brain to have an opportunity to share what it is thinking and feeling in relation to the topic in hand. Diversity of views makes for better business decisions as we have seen clearly in Chapter 8.

C for Control

Once each person has shared what's going on in their brain *leave the control with them to express their thoughts fully*. People need time to think and time to speak. Give them control of their own air time and have a commitment within the meeting that people will not be interrupted. This is very rare in my experience. Genuinely allowing individuals full control of their airtime to share and finish their thought is a high-quality way to generate ideas in a meeting and hear and feel what is going on for people in relation to the topic. Full brain potential is tapped into in this way.

A concern that executives have is that some people may hog airtime and others may feel intimated by this. Actually, this *is* what tends to happen in the average corporate meeting. In a brain-friendly meeting designed around this 4Cs model, if you are clear that you are creating the meeting context where everyone in the room has the time they need to express their thinking and that the expectation is that people are respectful in how this is

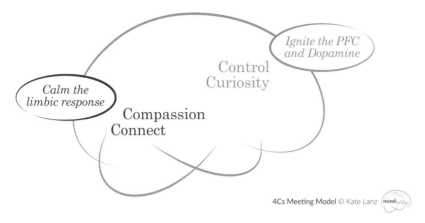

Fig. 9.2   4Cs Meeting Model™ (© Kate Lanz)

done then intelligent business people get the idea and proper turn-taking ensues in a smooth way. Limbic systems settle down and all the brains in the room feel much more relaxed about joining in. If they don't, then you are simply not running an effective meeting and you need to think carefully about your own skills in managing meetings and even, maybe, your own leadership style.

The other positive by-product of allowing each individual the airtime they need to share their thinking is that the overall quality of thinking improves as others in the room have the time and space to reflect and build on ideas that are coming forward (Fig. 9.2).

Case Study

A male client who led a team that was largely female was very keen to put into practice what he had learned about female-male brain differences. He had also had many experiences within the organisation of not feeling that his point of view was welcomed in meetings. He was an introverted, soft-ly-spoken sort and quite a lateral, reflective thinker. So he knew what it felt like not to feel truly included.

Armed with the new insights and information from the workshops he had attended on the differences between the male and female brain and within that the subtleties and nuance of understanding one's own brain sex, he decided to put to use the 4Cs Meetings Model™ with with his team.

He made four key changes:

- He cut down the number of items on the agenda.
- He framed the questions in relation to the agenda topics more broadly so they included how clients/others connected with the work were *feeling* about the work and he asked his team to consider issues from relational and emotional perspectives (as opposed to just getting the tasks done).
- He followed the 4Cs Meeting Model™.
- He made space for each person in the meeting to have all the airtime they needed with no constraints.

Energy levels were up. Participation and excitement were high. At the end of the meeting one of the women in his team commented, 'Wow – that felt different, in fact that was one of the best meetings we have ever had. What happened that brought about that change?'

Summary

The brain works bottom-up, through its emotional system into its cognitive system, resulting in our being responsive—via the embedded emotions—long before we have any conscious awareness of what our brain has already decided. Any form of productive work practice must take the brain's evolution into account and reflect this bottom-up approach.

Both of the RICH™ communication and the 4Cs Meeting Model™ take the brain's natural functioning into account. For the last few years I have been testing these models with busy executives in the hurly-burly of the modern business environment. They work. They create the context where all the brains in the business can feel validated, relaxed and creative.

The nuance of how each model is used will vary depending on the circumstance in which you find yourself. Are you running a meeting with a majority of high scoring male brained males? Or do you have a genuine mix of brains in the room? Or in a predominantly female environment what would you need to change and how? Taking the time to understand the likely thrive patterns of the brains that you are dealing with in any given circumstance will make a significant difference to how you deploy the models in practice.

Creating the conditions for all the brains in your business to not only come to the party but be asked to dance will improve your business performance very observably. Fact.

References

Eagleman, E., et al. (2018). Empathic neural responses predict group allegiance. *Frontiers in Human Neuroscience, 12,* Article 302.

Gawande, A. (2011, January). The Checklist Manifesto: How to get things right article. *Journal of Nursing Regulation, 1,* 64. https://doi.org/10.1016/s2155-8256(15)30310-0.

LeDoux, J. E., & Pine, D. S. (2016, September 9). Using neuroscience to help understand fear and anxiety: A two-system framework. M.D; ajp.psychiatryonline.org. Published Online. https://doi.org/10.1176/appi.ajp.2016.16030353.

Perlow, L. A., Hadley, C. N., & Eun, E. (2017, July–August). Stop the meeting madness. *Harvard Business Review, 95*(4), 62–69.

Rogelberg, S., Scott, C., & Kello, J. (2007). The science and fiction of meetings. *MIT Sloan Management Review, 48*(2), 18–21.

Wilensky, A. E., Schafe, G. E., & LeDoux, J. E. (2000, September 15). The Amygdala modulates memory consolidation of fear-motivated inhibitory avoidance learning but not classical fear conditioning. *Journal of Neuroscience, 20*(18), 7059–7066. https://doi.org/10.1523/JNEUROSCI.20-18-07059.2000.

10

Exhaustion, Energy and Excellence—The Male/Female Differences

Paul says:

The 3 Es

It has been central to the thinking in this book so far that emotions are the primary source of shaping the individual internal architecture of the brain in the first twenty-four years of life, and then using that structure to guide the rest of life. This happens because the human brain is designed in such a way that meaning is created by experience getting attached to the basic emotions combining into feelings which then all get described in language.

Embedded in the concept of 'emotion' is the intensely powerful variant, 'e-motion'. A memorable metaphor is provided by the famous consequence of Einstein's special relativity $E = mc^2$. Einstein evidenced that anything that has mass (substance) has energy embedded in it. Let's apply the metaphor to businesses. Organisations have mass, and energy walks in every day. It is what salaries buy in. The organisational dilemma is to be able not only to recognise what it is, and what the differences are as between women and men, but then how to get it to flow in the direction of the strategic and operational goals of the organisation because it is the application of human effort that transforms energy into profit. That profit then gets distributed in all kinds of ways through the multiple recipients called shareholders as energy residing in monetary value goes on flowing into the social system of which the profit-generating process was a part.

© The Author(s) 2020
K. Lanz and P. Brown, *All the Brains in the Business*, The Neuroscience of Business,
https://doi.org/10.1007/978-3-030-22153-9_10

Let's take E as meaning human energy. In our metaphor let us take the m of mass as being any one or any multiple of the individuals within an organisation. And then let's take c^2 as meaning 'care x capacity'. Now we have a memorable metaphor about human energy that is fundamental to the future of leadership's ability to access the full human potential in the business. This $E = mc^2$ metaphor then states that the available human energy inside the organisation, E, is a function of the way the organisation actively cares, c, for individuals and has the in-built capacity, c, to do so. We could just say that squaring c and making it c^2 makes it clear that an organisation requires a very great deal of care to get maximum energy.

My first ever visit to a factory was in 1959. As a second-year psychology student I had been required in the first summer vacation to make arrangements to visit any factory. It was part of the entry requirement for a second-year course on what was then called industrial psychology.

It happened fortunately that somewhere within my clergyman father's contact network was someone who worked at Carr's of Carlisle—the water-biscuit manufacturer whose products are still a delight to find in the minimarkets of S E Asia.

The person in question was called the Company Almoner—a title so unfamiliar today that, when tried as part of a brain re-engaging game after lunch on CPD programmes, it rarely elicits a correct answer. 'What is the modern title of what used to be called Company Almoner?', the question asks. 'Company Secretary' is the usual reply. But not so. Coming from the age-old practices of monasteries being the givers of alms to the poor, and the more recent practices of many pre-NHS hospitals where an almoner helped with a variety of social and financial problems that a patient might have, before social work took over, in mid-twentieth century organisations it was the forerunner of what used to be called personnel management, which then became HR, which now is looking for almost any other name as HR is the only part of large organisations that is continuously trying to find new ways of re-defining itself.[1] Lacking a scientific centre professionally, that's a very difficult task, as there is no agreed starting point from which a re-definition should stem. But this is not essentially about HR. It's about the certainty that what defines how much energy a person can make available to the organisation is a function of how much s/he feels valued by the organisation: and valuing people—not as assets; nor evaluating them; but

---

[1] *Harvard Business Review* has a new book due in June 2019 called *HBR's 10 Must Reads On Reinventing HR.*

for themselves—is a critical part of *care*: which is itself integral to the organisation's *capacity* to create the conditions under which energy flows in the required directions.

***

Synthetic or Authentic?

Enough abstracting, for the moment. How does it work in practice?

If, as seems irrefutably the case, the quality of relationship within which any one of us exists is profoundly important to our sense of well-being, that state of well-being is a way of knowing that energy is comfortably available to be directed in any appropriate direction. The more that the attachment emotions are active, the more energy there will be to flow. But that depends on the integrity of relationship.

Synthetic emotions don't do it, of course. Nor do polite expressions of the 'right' emotion. In 2005 the then Archbishop of Canterbury, Dr. Rowan Williams, apologised for the Crusades. In 2008 he also apologised to Charles Darwin for that fact that the Church of England had ever doubted Darwin's theory of evolution.

Leaving aside the whole question of whether history should be revised in such a way and whether acts of such kind have any meaning at all, what was especially curious from a human point of view about both those apologies was that there was no-one on the receiving end with the capacity to consider the apology and, if accepted, to properly forgive. It needs a working relationship for an apology offered to be effectively received.

That is an illustration of the sense that this concept of 'capacity' in our re-defined $E = mc^2$ metaphor only has meaning when it is made evident in the context of active relationship. Such an observation leads to the idea that part of the difficulty that modern leadership is that it has so abstracted, some profoundly important aspects of human endeavour that they have not only lost effective meaning but waste huge amounts of energy organisationally: energy that is not therefore available for the main purposes of the organisation.

In current organisational practice, the words 'diversity' and 'engagement' have got themselves thoroughly embedded: and in many large organisations both have not insubstantial resources attached to them. Not infrequently 'women's issues' are siphoned off into diversity. When one looks at what happens operationally, 'diversity' seems to be a convenient catch-all into which anything that is slightly inconvenient about differences in attitudes, behaviour, or social values can be siphoned off, perhaps then to be addressed with the pernicious processes of tackling 'unconscious biases'.

The harm that is done by an abstract concept like 'diversity' is that it stops serious consideration of the value of the *differences*, which is what 'diversity' really means.

And something similar happens with 'engagement'.

Considerable amounts of data now describe how alienated many—often more than 50% of employees—feel at work. But the very survey method by which their alienation is assessed it itself an alienating process. In *Forbes Magazine* in September 2016, Liz Ryan describes simply and elegantly how to get effective communication working. It relies on proper listening more than messaging, and is not well done by impersonal processes.

The point at issue is that, without a central understanding of what 'a person' is, modern leadership, including HR, struggles to create processes that are a reflection of a such an understanding: and that makes it very easy to create, by default, the conditions under which adverse effects happen. Like burn-out.

When did performance-driven cultures become the western-world's favourite organisational paradigm? Moving from a rear-mirror perspective for moving forward to a forward-looking, goal-oriented approach must have seemed such a good idea, put like that.

But twelve years ago Chan Lismen (2004) and his colleagues could find no evidence that high-performance human resource practices actually produced the results they were avowedly supposed to produce: though as recently as 2017 Nicole Morgenstern was promoting performance-driven cultures for the American Management Association through its Playbook series.

We know that management proceeds on a very pragmatic basis. 'Tell me what works, not why it works', is the demand of most managers. And HR is part of that culture. But trial and error is not a way to proceed when there is knowledge to be applied: and especially not when human beings are at risk.

The Burn-Out Risk Comes About This Way

In the first place there is a complete lack of understanding about what 'a person' is. Twentieth-century psychology developed lots of theories, but no shared understanding. In claiming to be the science of human behaviour or the science of the mind it served HR very badly indeed.

Secondly, 'energy' as a centralising concept applied to humans at work has been, with one exception, entirely missing from management textbooks, management theory and psychology textbooks. The one thought-through exception is a proposal by Lex Sisney in 2012 that the fundamental concepts of physics should be applied to organisational thinking and design.

And thirdly, the concept of being performance-driven seems not to have had any sense of how to regulate the processes of getting to whatever stretch goals have been set. There has been a great deal about how to measure performance, but nothing to balance that on how to conserve, renew, or even create the energy that would be appropriate to the performance demands.

So performance-driven cultures run the risk of being like a car that is driven in top gear under all conditions. When the hill gets too steep, the foot goes harder on the accelerator. Steeper still, and the car judders to a halt. That's a breakdown. Or, in trying to manage things as best s/he can, a performance-driven executive gets more and more exhausted but accepts that as a way of life. Sleep deteriorates, skin gets pallid, eyes begin to have no sparkle and in women especially hair starts to thin.

What is happening inside is that the normal bodily processes that produce adrenalin for energy and cortisol to focus attention and manage sleep begin to get out of kilter. Something called the HPA axis, which is the main controller for all stress responses, goes into a loop in which it cannot recover its normal function. So though the body is eminently well-designed for dealing with stress, what it cannot cope with is continuous stress. It's like running a marathon, and then another, and then another with no recovery time. Under continuous stress the body loses its capacity to regulate itself properly. Then the system's only way of signalling that it is in distress is to force recovery time via burn out. That's actually very adaptive, but in the Western world we have been taught to think of it as illness.

So unhappily we have come to regard the burn-out or breakdown itself as a disorder, not the body doing its very best to protect us by creating the conditions under which it might have a chance to recover. As with pain, which we tend to regard as the disorder when in fact the body is signalling to us is the only way it can, having no language of its own other than bodily sensation, that it is tackling something that is demanding its resources and is asking for help. Epigenetics is beginning to give us a new fix on such ideas.

The American Psychological Association (APA 2012) has a useful summary of the differences between men and women in response to stress, and lays some emphasis on the differences between married and single women though, interestingly, offers no comparative data on single and married men.

Surveying over 6000 people across four years, an overall summary conclusion is that:

> Men and women report different reactions to stress, both physically and mentally. They attempt to manage stress in very different ways and also perceive their ability to do so — and the things that stand in their way — in markedly different ways. Findings suggest that while women are more likely to report physical symptoms associated with stress, they are doing a better job connecting with others in their lives and, at times, these connections are important to their stress management strategies.

The common assumption is that under stress men go for fight or flight—become aggressive or withdraw. Women go for contact in relationships—tend and befriend. The APA data lends some credence to this.

When the whole organisation is under stress, the danger is that the men will fight or disengage. It is a product of feminine energy to use the generating power of relationships to sort out whatever needs sorting out. In doing so, women run up against the problem that men want an answer, construing everything as a problem to be solved with an outcome to be given. But the real power of relationship is that it doesn't just sort out *this* problem: it creates resource at the same time for the next and the next and the next. That creates the possibility of continuous excellence, and it comes from e-quality.

References

American Psychological Association. (2012). *Gender and stress.* https://www.apa.org/news/press/releases/stress/2010/gender-stress.

Lismen, L. M. C., Shaffer, M. A., & Snape, E. (2004). In search of sustained competitive advantage: The impact of organizational culture, competitive strategy and human resource management practices on firm performance. *The International Journal of Human Resource Management, 15*(1), 17–35.

Morgenstern, N. (2017, December 13). *Cultivating the elements of a performance-driven culture.* American Management Association Playbook. https://playbook.amanet.org/training-articles-performance-driven-culture/.

Ryan, L. (2016, September 18). Ditch the employee engagement survey—Here are ten better ways to listen. *Forbes Magazine.* https://www.forbes.com/sites/lizryan/2016/09/18/ditch-the-employee-engagement-survey-here-are-ten-better-ways-to-listen/#616c9f827793.

Sisney, L. (2012). *Organizational physics: The science of growing a business.* Kindle edition. https://www.amazon.com/Organizational-Physics-Science-Growing-Business/dp/1300785632.

11

Creating the Mind of the Organisation

Paul says:

The Courage to Go On

Someone inside the organisation has to take on the extraordinary and huge task of making the organisation fit for women. The pressures and demands can come from below and outside but, being structured hierarchically as they are, effective action is going to have to be not only sanctioned but imaginatively encouraged from the top.

That word 'encourage' is worth dissecting, for a moment. Like 'emotion' containing within it 'e-motion', 'encourage' offers the opportunity for an interesting discovery.

Many years ago I was listening to a clinical colleague talk about his work with elderly people. In the Q+A, someone asked the question: 'What would you say is the single best thing that can be done for an elderly person in care?'

'Do whatever is necessary to help an old person find the courage to go on', came the answer. And then came one of those flashes of thought that stick forever. 'Among our clinical group we have started using the French way of saying that word. What we are trying to do is what we now call en-courage the elderly. Say that as if speaking French. It gives a completely new emphasis to the English word "encourage"'.

© The Author(s) 2020
K. Lanz and P. Brown, *All the Brains in the Business*, The Neuroscience of Business,
https://doi.org/10.1007/978-3-030-22153-9_11

Courage is one of the five special qualities of the leader who engages the brains and belief of others. It's a topic that we explore more in the next and final chapter. But before asking how today's leadership including the HR function is going to reinvent itself to be the agency that activates that courage, let's get back to the brain and what it might tell us about the development of a twenty-first century organisation that is both sustainable and where energy flows for the satisfaction that purposeful work creates.

Among the many extraordinary things about the brain, one that goes unremarked most of the time because we take it so for granted is how the brain takes the data from the six sources it takes data from and puts it all together seamlessly: and then makes meaning and attaches that to language and presents it to the brain's owner and the outside world as if it were all quite simply done and not an extraordinarily complex process at all.

Five senses give the data from the external world. A variety of perceptions—thoughts and feelings especially—come continuously from the internal world. The brain takes them all as data and integrates them without us having any awareness of any kind at all of the journey that any bit of data has been through in the process of combining with all other relevant bits of data.

We know that the brain has some highly specialised areas. The visual cortex is located at the back of the brain, for instance, and processes everything we see before it gets attached to memory—and does that at lightning speed. Everything connected with body posture, walking, never forgetting once learnt how to ride a bike or swim, is located in the cerebellum at the very base and back of the brain. That's only two areas. There are three more major areas and thousands of way stations in the integrating of information. Connectomics is the new science that is working out how the brain does put it all together so seamlessly. It's the fact that it does that's important to us here, though.

Think of any organisation of which you know. Think of its main departments. Then think of the boundaries between them and how much effort and energy goes into defending and maintaining the boundaries. If the brain were doing that at the boundaries of its specialised areas it could not function as a whole system with all its specialist departments. In the brain they are called 'regions', but are just the same as organisational departments.

So why is it so difficult for departments to operate all the time as part of a whole integrated system? Why do people fight across the boundaries and waste huge amounts of otherwise productive energy in doing so?

There Are n Interlinking Answers, and Some Are Here

The first is because modern organisations have established themselves as expressions of the way men organise their worlds. So in coming into organisations in their own right, women have unquestioningly or perhaps unavoidably adopted men's rules.

After sitting through a board-room meeting full of conflict, Angela said in exasperation as we started a coaching session: 'Why don't they *get* it?' For her, the unproductive waste of time in taking positions was such an obvious waste of corporate resources. Challenged by being asked what could she have done to change the meeting, her immediate reply was to the effect that it would be like a man in a rubber dinghy trying to push against the prow of an oil tanker to get it to change its course. 'No matter how powerful the engine', she said, pointing to her feisty self, 'it wouldn't make one bit of difference'.

'But that's playing their game', I said. 'What would a woman do?'

'Ah', she said. 'We women haven't worked that one out yet'.

'Cop out', I said, and we both laughed, but the challenge stayed hanging in the air. After a little quiet, Angela said: 'It's got to be different, hasn't it? Could we do it from the bottom up?' And she started a conversation about whether her organisation could possibly consider the idea that it might set up, quite deliberately and clearly, a business unit staffed only by women; with a brief to experiment for three years with how women would do it differently and what implications it would have for existing organisational systems.[1]

We then talked about the mid-terms issue that would arise, reflecting on a professional experience of some years ago when the Chief Executive of a large local Authority in the UK had called asking if I would come into a discussion about a problem they had and did not know what to do next.

It transpired that the Authority had for some time been concerned about the rising costs of dealing with problem families in its area. Some in-house auditing had come up with the observation that about a hundred families were costing the Authority ten million pounds every year in direct and indirect costs.

---

[1]At the time of writing it is not known what the organisation will decide, but there is an active long-range policy discussion taking place around the question and the idea is being taken seriously.

To find a sustainable answer the Chief Executive had, two years previously, set up an experimental unit staffed by a small consulting team from outside the authority and whose leader was an Australian woman. The consultancy had a reputation for understanding the practical applications of attachment theory and what makes for families and relationships that work well.

The essential brief to the consultancy had been to do whatever they considered professionally responsible in the interests of seeing whether, over two years, they could make a real difference to the ways that six of the problem families functioned. It was a reciprocal requirement of the consultancy that when, as they expected would not infrequently be the case, they called on the resources of the Authority for whatever help they needed, and in particular the resources of social services, education, police and fire brigade, those resources would act as the consultancy requested of them whether or not that request was within the normal operating rules of the service in question.

It doesn't take a lot of imagination to work out what personal time and effort the chief executive had to put into getting the Heads of Services to agree to such a requirement. But they all understood the problem because all their services were variously impacted by it. It so happens too that professionals in those disciplines profoundly dislike having to gear up to cope with the kinds of repeated crises that problem families can create in the knowledge that whatever is done is a short-term fix and will not have solved a fundamental problem; and that there will be some kind of repetition of the problem they have been called in to deal with on a crisis-driven basis.

Paradoxically the fact that the services *were* available to problem families when in crisis was part of the only stability in the problem families' world. A crisis brought to them the coping skills of others that were missing from within their own family units. Not that this fact would have been recognised as such, for there would often be a sense of aggressive entitlement in their demands upon whatever short-term help arrived. And of course there was never any lasting benefit, just a worse crisis averted and logged as all in a day's work until the next made its own demands again.

I was called into the discussion about what to do next when, after two years, the experimental work had been profoundly successful. Members of the consultancy team had virtually lived with the families concerned, and bit-by-bit had helped them develop quite different relationship skills. The annual costs to the authority had moved from more than a million-and-a-half pounds per family per year to ten thousand pounds. The experiment had more than paid for itself and there would be continuous gains both financial and operational for years ahead, let alone the long-term next-generation benefits for the children involved.

Winston Churchill once observed that it was not difficult to decide what to do. The real problem was what to do next. And that was exactly the Authority's dilemma. What the success of the project had shown was how much the main services in the authority were geared up for crisis, not prevention or long-term remediation. More importantly, the requirement of the consultancy that services respond out of need, not out of protocol, had shown how many protocols had developed in the various services that were designed to fit the needs of the service, not the needs of the situation. The problem about what to do next focused on the question of whether it was feasible for the Authority to consider a wide-scale, radical change of its operating style into being citizen-focused not process focused.

'That', I had observed to the meeting considering the dilemma, 'is what the experiment you are having to think about will challenge the organisation with. It's not just "can we do it differently?" You know the ways you can do it differently. You have demonstrated them to yourselves over the past two years with all the benefits resulting that you have seen, but how does that get embedded into the organisation?'

What had been discovered was fundamentally about the nature of the way the brain works: co-operative relationships at all levels in its organisation. The Authority was learning about itself, not just about an enormously successful cost-reduction exercise with the families involved. The brain is systemic. It functions continuously as a whole system. But if an organisation is going to be sustainable and integrated in the same way, it has to take everyone with it. And that's not easy when men tend to go for either/or outcomes and women go for both/and solutions. The way through is, in its essence, not about compromise, though in practice of course it sometimes is. It's about seeing all the parts interlocking and have mutually dependent effects. And linking everything together. That's what the brain is so good at in managing itself.

After two years and serious success, the question in the Authority that demanded an answer was: 'How are you going to get that pattern of mutually supportive relationships embedded throughout the organisation?' Protocols won't do. They create mindless organisations—organisations that mind first of all about process, not people. An organisation that is the opposite of 'mindless' is not 'mindful'. That has become a specialist word for practices that let the brain start to settle itself into a state where any subsequent action is possible. The opposite of being mindless is, simply, *minding*. It's the active verb that underpins our c^2—care and capacity.

***

But We Don't Really Know What 'an Organisation' Is

Which raises the question; 'What is "the organisation?"'—this setting to which so many people give the greater part of their lives: that at law has an independent identity such that its shareholders and directors are required first of all to service 'its' interests: whose demands can destroy lives, make fortunes, be intensely disruptive, be hugely successful and have remarkable failures. If women are going to make a difference to 'it', what is 'it'?

The fact is that there is absolutely no agreement about what 'an organisation' is. There is no agreed working model. Like psychological theories about the individual, but no agreed definition among psychologists and psychotherapists as to what 'the person' or 'the Self' actually *is*, so organisational theory has failed to come up with any agreement as to what 'an organisation' is.

In 1983, Gareth Morgan produced what became a classic book called *Images of Organization*. Its impact was to draw to business schools' attention the fact that organisations seemed to fit within a variety of metaphors. He suggested seven in all. It proved to be great stuff for teaching. It is the epitome of an observational descriptive style of considering organisations theoretically. Ten years later Charles Handy (1993) did something not dissimilar with *Understanding Organisations*: and in 2016 Ortenblad, Trehan and Putnam (2016) edited a volume *Exploring Morgan's Metaphors: Theory, Research and Practice in Organizational Studies.* Type Morgan's Organizational Metaphors into Google and it can quickly be seen what an extraordinary academic industry there is around describing organisations.

Description, however, tells us only how they look, not how they work. And metaphor only tells us something about how an organisation works by reference to something else whose working may or may not be understood.

Nearly half a century on, organisational theory has no more come up with an agreed working definition of what an organisation is than psychologists have for understanding people. So whatever it is that women might do to re-invent organisations, a pre-requisite might be that there is some understanding of what it really is that they are setting out to reinvent. It won't work to know what one *doesn't* want. Imagine planning on going on holiday by saying where one didn't want to go to. It would be difficult to start the journey, let alone have bought the tickets or pack for whatever the climate might be or the fun to be had.

Laloux (2014) has given some practical pointers in his detailed descriptions of the organisations he has found that do appear to have reinvented themselves. In his 2014 *Reinventing Organisations* he describes organisations where authority is kept as low down the organisation as possible, wherein consequence responsibility is high, where the quality of relationships engenders high levels of trust, and where the central purpose of the organisation are what its members are willing to give a large chunk of their lives for. He has recorded how organisations that are relationship-driven rather than performance-driven produce outstanding sustainable results *not* because profit is the purpose of the organisation but because organisations that are purposeful also produce profit.

Laloux's fieldwork is not underpinned by anything neuroscientific. It was not his frame of reference. What is profoundly important is that there are beginning to be worked examples of organisations that appear to run on what might be called feminine principles.

But that still doesn't get around the question of how is one going to think and talk about 'the organisation' that one wants to be different when there is no agreement as to what 'an organisation' actually is?

It has to be confessed that metaphors are useful if put into use. So here is one that has huge implications for thinking about organisations. It starts from the way medicine has developed over the past two-and-a-half centuries.

Go back to the first half of the eighteenth century. There was no agreement among physicians as to how the body worked: and physicians despised surgeons because surgeons did not study at university but learned their trade rather as butchers' apprentices did, on the body. Surgeons had also been closely allied to barbers for two centuries, since Henry VIII granted a Charter to the Honourable Company of the Barber Surgeons in 1540. With the fashion for huge wigs that gained serious momentum when Charles II came to the throne in 1660 and continued unabated until Prime Minister William Pitt the Younger put a tax on them in 1795, surgeons became increasingly disaffected at being identified with barbers, and by 1800 acquired their own Charter that is the origin of the Royal College of Surgeons of London.

Another major change was happening in medicine, though. It was becoming scientifically-based. The great triumph of western medicine has been to develop itself from the experimental sciences that systematically set out to *understand* what is going on not just *describe* it. Organisational theory has not got there yet. It is still like eighteenth-century medicine. One organisational healer, a consultant, suggests this: another suggests that. Large consultancies have their own nostrums that sometimes work and sometimes don't, but the patient pays anyway.

So the useful thing for organisational practice would be to insist that the business schools start doing what medicine did, which was to agree what the vital organs are, begin to understand what each one actually does, then how each affects the other as part of a whole system, and what it is that links them all together.

It won't at this stage be a surprise to know that, from a neuroscientific perspective, it is the flow of human energy that links all the vital organs of the organisation together. And if, as we are asserting throughout this book, the way female and male energy presents itself and flows is essentially different then the clear conclusion is that an organisation run on feminine as against masculine principles would be different. The female brain can be just as interested in profit as the male brain. So the purposes for which male and female energy may be used can be identical; but the means of getting there would be different.

References

Handy, C. (1993). *Understanding organisations.* New York: Oxford University Press.
Laloux, F. (2014). *Reinventing organizations.* Norton Parker.
Morgan, G. (1983). *Images of organization.* Newbury Park, CA: Sage.
Ortenblad, A., Trehan, K., & Putnam, L. (Eds.). (2016). *Exploring Morgan's metaphors: Theory, practice and research in organizational studies.* London: Sage.

12

The New Organisational Paradigm

Paul says:

Getting Towards the Kind of Organisation Fit for Women and Men Together

So now we have some elements for the imaginative re-design of the twenty-first century sustainable organisation that is fit for the purposes of women as well as purposed to fit the men. For the first time in organisational theory and design it begins to be possible to see that 'the organisation' and those who people it have a single common origin and purpose: they are energy systems, and the practical question is how to turn human energy into profit. And the operating consequence from that observation is how to design the organisation so that all the brains within it, differently powered as they are for different purposes in life, can be used to maximum benefit for both the individual and the organisation.

'.. how to turn human energy into profit' seems such an obvious formulation of what any organisation is about, whether it's financial profit or social profit as in the work of charities or the business of local or central government. And yet, to the best of our knowledge, it has never been formulated before, quite like that.

In Chapter 10 we made some reference to Lex Sisney's thinking about how to apply the fundamental principles of physics and the laws of thermodynamics to growing a business. But in looking for general principles what

© The Author(s) 2020

K. Lanz and P. Brown, *All the Brains in the Business*, The Neuroscience of Business, https://doi.org/10.1007/978-3-030-22153-9_12

his creative thinking misses out on is the individual complexity of the m part of our $E = mc^2$ metaphor. What we are proposing here is the *beginnings* of a general theory of the organisation and the individual. The starting point is not the organisation but the differences and complexities of the individuals who are the living and effective presence of 'the organisation'.

With only a little hindsight we can now clearly see that in twentieth-century psychology's remarkable failure to offer a coherent and agreed working model of what 'a person' actually is lie the roots of modern leadership's reliance on abstracted ideas like 'diversity' and 'engagement' that generate processes that in turn and in combination bear no effective relationship to what any human being understands about her—or himself. So the process has been imposed on a person. That is perhaps why there is such a corrosive underlying sense of alienation between individuals, modern leadership and the organisation.

***

At Last, a Working Model of the Mind

Before we put the elements of the new organisational design together, there is one last major understanding of the way the brain works that needs to be in play. And that is about the nature of mind to which there was the beginning of a reference in the last chapter.

Among many things that psychology has neglected to tackle experimentally, or even form an agreed view upon, is 'mind'. Despite a common definition of psychology being a body of knowledge about 'the science of the mind', that elusive appurtenance of the human condition has been a subject more for debate than discovery and the province of philosophers rather than experimental psychologists.

However, in 2010 at a conference in Boston, Dr. Dan Siegel—psychiatrist and a world thought-leader and practitioner in how attachment theory can be put widely into use in a reformulation he calls 'interpersonal neurobiology'—offered the practical idea that we can see a mind at work in ourselves and others in the outputs of the way that the brain is continuously managing information, energy and relationship, all three in a dynamic relationship one with another.

It is the modern neurosciences that have made such a formulation possible and it has exciting practical consequences individually and organisationally. Think of this. For the last thirty years at least individuals in organisations have been overwhelmed with information, as organisations have maxed on their ability to produce and distribute it. But there has been no parallel corporate understanding or managing of how energy gets wasted with individuals exhausted by being overwhelmed; or how that energy needs systematic replenishing and proper management. Machines get planned maintenance, human beings not at all.

And relationships have been increasingly defined in protocols of behaviour that require both conscious energy to manage them and have anxiety about failing to meet the protocol's requirements connected to them. In the implied low trust that is contained in many human process interventions in organisations, adults are told how to behave as if they were in some kind of kindergarten.

Figure 12.1 gives a simple if slightly static and two-dimensional way of conceptualising the working of the mind.

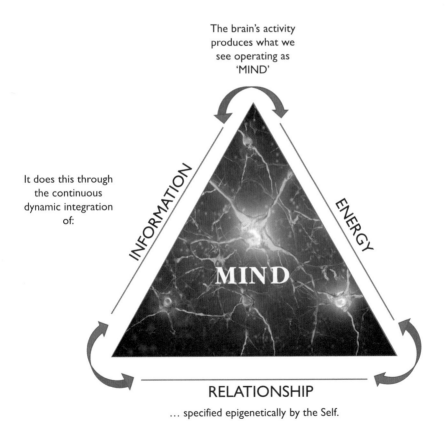

The brain's activity produces what we see operating as 'MIND'

It does this through the continuous dynamic integration of:

INFORMATION

ENERGY

MIND

RELATIONSHIP

… specified epigenetically by the Self.

Fig. 12.1 Conceptualising the working mind (adapted from Dr. Dan Siegel)

This is what a modern HR function could re-organise itself around. The special benefit of doing so would be that, properly done, the organisation would be continuously experienced as *minding*. This is the opposite of the mindless nature of so much that happens with protocols of behaviour. After all, protocols are produced by some other mind or minds. How could they possibly fit all other minds in all circumstances except by so defining the importance of the protocol that all other considerations have to give way before it, as happens with potentates and dictators.

Real Life Mindlessness at Work

Last year I changed my mobile phone provider. It meant going through the routine of shifting the number from one supplier to another. The final bill was duly paid with some difficulty as it came after the account was closed so that on-line payment access was barred. But then a month later a new bill arrived from the former supplier for a trivial amount—less than five pounds—without any explanation of why. It seemed that even at six thousand miles distance a conversation with a real live human being might be the quickest way of sorting it out, and instructions were available for how to do that.

The transcript record of the messaged conversation that lasted from 9.33 to 10.53 p.m. has, just after twenty-minutes of protocol questions and what appear to be unsatisfactory answers, an observation from the member of the team that says:

> *I already have an answer to the question you have asked.*

Paul replies:

> *I should be delighted to have it. Thank you.*

The company representative replies:

> *Till the time the security check is not completed, we are not allowed to give any details of the account. I understand it's frustrating for you.*

Not wishing to complicate the exchange with a discussion about quite what the first part of the sentence actually meant, Paul pressed on. By 10.10 p.m. the team member's only option was to tell him that he had to go to the nearest local representative outlet to sort the matter out, and an address for the store in Royal Tunbridge Wells, Kent, appeared. Paul typed, in reply:

That is 6,000 miles away from where I live. What course of action do you propose?

She says:

I totally understand your concern as well.

And then immediately adds:

If this would have been in my hand, I would have helped you.

Paul writes:

…. while it is kind of you to understand my concern, what effective action is to be taken? Your company is billing me for monies to which, as far as I know, it is not entitled. I have now spent 45 minutes answering mindless questions because of the system. There must be a human being in your company who has the authority to take effective action. Who is that person please?

She responds:

Paul, I completely understand your reasoning for the situation, however we have to follow a strict security protocol …. Please accept my apology for that.

There is another three-quarters-of-an-hour of exchanges in the same vein. Enough is here, though, to demonstrate what mindlessness is about; and to hear the attempts made to create some kind of understanding relationship within the rigid confines of a requirement of a protocol. She had no authority to manifest either care or her own apparent capacity and the joint waste of energy was profound.

Frederic Laloux's *Reinventing Organisations* (Chapter 11) is especially thoughtful on the question of whether or not an organisation trusts its employees: trusts them to be self-directing systems who, if the organisation has purpose and creates freedom to act responsibly, manifest the best that human beings in shared endeavour can produce.

He also makes interesting observations on what happens when an inter-personal virus gets into the system—a person who feeds off the system, rather than contributes, and who absorbs energy rather than exchanges it. A healthy system gets rid of the virus and repairs the damage that has been done, and it does that as part of its capacity to do so. So if appropriate care and proper capacity create an organisation that minds: and an organisation that minds goes on increasing its capacity to do so, then like so many biological functions, there is a circularity in the system that, properly functioning, reinforces itself. And at its best, the emotions that flow are all varieties of trust and many varieties of love.

The Sustainable Organisation of the Twenty-First Century

So now we can specify the elements from which the new organisational design can be constructed.

The first is understanding energy, and the different ways women and men deploy their energies while not losing sight of the fact that they can be equally interested in the outcomes.

The second is developing the organisation's capacity to care continuously, appropriately and effectively: with the recognition that that is a more natural capacity of the feminine brain than a masculine trait.

The third is the awareness that effective relationship underpins everything because of it both an expression of and a generator of trust. And trust is the emotion that lets the brain function as its most flexible and agile.

The fourth is to construct a working model of the organisation that makes it possible to map and track individual and corporate energy flow, for that will define the well-being of the whole system.

The fifth is that information is a curious commodity, as an illustration will shortly show.

The sixth, more useful to have in the background of one's awareness than very upfront, is that the brain belongs to its owner and the job of leaders and managers is to create the conditions under which everyone's brain wants to engage with the composite organisational brain. Like the players in an orchestra tuning to the leader of the orchestra and giving of their best to a conductor who inspires them into giving of their best, so any organisation can create similar circumstances. And, like the trumpet players when there is a long gap before they will be in play, it's alright to sneak off and have a smoke.

The seventh is that women and men are in it, whatever 'it' is by way of the organisational life they are in, for very different biological reasons that require social expression of the differences if women's energy is going to really come into the organisation. For men 'it' tells them, hopefully, who they are. Men know themselves through striving. Women know themselves through expressing themselves and seeing a continuous process of finding good solutions not, as for men, being mostly focused on solving problems.

Eighth and last is the avoiding the terribly easy trap of creating a mindless organisation, busy with keeping its systems going but completely dysfunctional so far as trust is concerned. Such organisations sap individual energy and marginalise individual productivity and the kind of satisfaction of belonging to something that is so worthwhile it makes going to work one of the central satisfactions of a life.

And What Would the Capacity of the Leader Be Like?

What would the kind of leader be like who created such conditions?

On a long Saturday afternoon's mind-mapping exercise ten years ago now, in preparation for a lecture at the Royal College of Defence Studies' year-long programme on global strategic leadership, I set out to synthesise what was in my head and on the bookshelves about leadership. What, I wondered, were the irreducible characteristics of the kind of leader who others wanted to follow.

Five qualities appeared, and the leader who evidences them might be called a limbic leader because s/he is operating on the basis of intelligent emotions.

'Intelligent emotions' are distinct from emotional intelligence, which is a concept that has allowed itself to become degraded through being measured as if it came with measurable quantities attached to it, like cholesterol. Intelligent emotions describe the process in which an individual can both name, value and perceive their own emotions in the immediate context in which they are occurring, acting as the main radar system through the amygdala for assessing all external and internal signals, and then can attach those emotions seamlessly to thought and expression of the feelings arising from the emotions, embedded in the language that thought generates.

The five qualities that coalesced out of the mind-mapping exercise, all of which are individual capacities, were:

Being able to connect

Being courageous

Being clever enough

Walking one's own talk

Inspiring others into action

The flash of discovery that the arose from that list when it finally appeared was that each of those five qualities relies on the person trusting him or herself. And when they are evident in a leader, then s/he is trusted. Others' brains tune without difficulty to a trusted leader's brain and do their best to create harmony.

When Algorithms Take Over from People

With the advent of algorithms, trading in all sorts of financial instruments has changed dramatically. A highly experienced trader looking at complex patterns continuously on a screen, and with years of data stored non-consciously until needed, can make critical buy/sell/hold judgements about every three seconds. It takes a lot of brain power and is remarkably exhausting, which is why that industry used to be noted for its evening social excesses by way of recovery. But with the advent of mathematicians creating algorithms, not only have the excesses somewhat disappeared along with the desk traders but the trading decisions can now be made, while tracking what all competitors are doing at the same time, at the unimaginably fast speed of well over one-hundred-thousand *per second*.

That kind of disruption demands wide-ranging wondering about how then to structure a company. If competitive advantage no longer resides in hiring the right traders, and relatively few but intensely clever mathematicians can create algorithms that may vary between one- and two-hundred thousand decisions a second, all competitive advantage as previously known is lost. What is a company to do to retain its competitive advantage?

One relatively small, specialist trading company in one of the financial capitals of the world decided that the quality of the interpersonal environment—relationship—would be the key differentiator. So over three years the Chairman and CEO gradually felt their way to creating an organisation to which staff *wanted* to come.

Among many difficult decisions about who and what was right for the organisation, one was to show complete trust in every member of staff by having one room in which all company information was continuously displayed and updated. Nothing that was known to the CEO was not displayed. There were only two rules. All information was confidential but could be discussed within the company by any members of staff wanting to discuss it together; and any questions about it could be asked at the regular Friday morning shake-out staff meeting.

During the three years, there was one very difficult set of circumstances where it was felt a serious breach of trust had occurred involving three senior people. In the legal processes that followed, the barrister eventually observed that he had learned a surprising amount about how to conduct difficult negotiations when the company's continuous reference point was whether what action was to be taken fitted with the company's insistence that trust was at the heart of the way it worked, even in the most difficult circumstances.

And one surprisingly gratifying and unexpected outcome of what was being attempted was that at the end of the first two years there was a surprise request from the Financial Regulatory Authority to visit and talk with the firm about what was making the buzz in the market place they were hearing about. The firm had become the go-to place for the brightest and best wanting to be in a setting where the quality of the everyday experience of being at work was as good as it could possibly be. Happily, too, trading results were better than ever. And all this happened in circumstances where people were given masses of information but able to access it and understand it from within their own needs. So it may be that it is not just the *quantity* of information that overwhelms and stresses people in modern corporate life but the *context* within which they have to worry about using it.

There is a rippling systemic effect in a working environment where people feeling known as individuals underpin all relationships and all decisions. In 'The Check-List Manifesto' that was referenced in Chapter 9, trauma surgeon Atul Gawande observed it in a particularly unexpected way.

You will recall that he had come to the conclusion that there was something absent in the quality of personal encounter within the trauma surgical team: that however well it functioned technically it was *im*personal. So he built into the very start of the team getting into theatre, ready for action, a requirement that the members of the team introduce themselves one to another.

The ripple effect he especially observed was that patients who had been operated on by a team whose members knew each other by name stayed in hospital on average one day less than patients whose surgical team had known each other only by function. So he set out to explore his conclusions nationally and internationally, in widely different settings, and found them replicated with remarkable consistency. But, through a carefully constructed questionnaire study, what he also found was that some surgeons who didn't want to be bothered with bringing a sense of the personal into theatre would nevertheless, should circumstances arise where they themselves were on the operating table, prefer to be operated on by a team that had introduced itself to itself.

In Conclusion—All the Brains in the Business

Kate and Paul say:

The brain is the organ of relationship. Relationship adds something well beyond artificial intelligence or technical skill. The quality of relationships and the trust engendered through them, is one of the primary ways that human energy moves through an organisation. Thus relationship impacts how creative and agile an organisation is and how effectively it can execute its task.

Businesses that know the value of female brains as well as those of men and how to create the trusting relationships that enable access to brain gender difference (in all its diversity) will win. Some divisions in big business and some smaller companies are beginning to wake up to the size of this prize and are already noticing the benefits. We have shared our experiences, so far, of this with you in this book.

In the interests of enabling organisations to support sustainable futures for human health and well-being and the planet we all inhabit, we hope this book will inspire you to make positive change for yourself and all the brains in your business.

Reference

Gawande, A. (2009). *The check-list manifesto: How to get things right*. New York: Metropolitan Books/Henry Holt and Company LLC.

Further Reading

Bachevalier, J., Brickson, M., Hagger, C., & Mortimer, M. (1990). Age and sex differences in the effects of selective temporal lobe lesion on the formation of visual discrimination habits in rhesus monkeys (Macaca mulatta). *Behavioural Neuroscience, 104*(6), 885–899.

Lipton, B. (2006). *The wisdom of your cells* [Audio CD]. Louisville, CO: Sounds True Inc.

Lipton, B. (2015). *The biology of belief: Unleashing the power of consciousness, matter and miracles.* London: Hay House UK Ltd.

Index

© The Editor(s) (if applicable) and The Author(s),
under exclusive license to Springer Nature Switzerland AG 2020
K. Lanz and P. Brown, *All the Brains in the Business*, The Neuroscience of Business,
https://doi.org/10.1007/978-3-030-22153-9